DEVELOPMENTS IN PETROLEUM ENGINEERING—1

THE DEVELOPMENTS SERIES

Developments in many fields of science and technology occur at such a pace that frequently there is a long delay before information about them becomes available and usually it is inconveniently scattered among several journals.

Developments Series books overcome these disadvantages by bringing together within one cover papers dealing with the latest trends and developments in a specific field of study and publishing them rapidly.

Many subjects are covered by the series including food science and technology, polymer science, civil and public health engineering, pressure vessels, composite materials, concrete, building science, petroleum technology, geology, etc.

Information on other titles in the series will gladly be sent on application to the publisher.

DEVELOPMENTS IN PETROLEUM ENGINEERING—1

Edited by

R. A. DAWE

Petroleum Engineering Section,
Department of Mineral Resources Engineering, Royal School of Mines,
Imperial College of Science and Technology, London, UK

and

D. C. WILSON

ERC Energy Resource Consultants Ltd, London, UK

CRC Press
Taylor & Francis Group
Boca Raton London New York

CRC Press is an imprint of the
Taylor & Francis Group, an **informa** business
A TAYLOR & FRANCIS BOOK

CRC Press
Taylor & Francis Group
6000 Broken Sound Parkway NW, Suite 300
Boca Raton, FL 33487-2742

ISBN-13: 978-0-853-34358-5 (hbk)

Publisher's Note

The publisher has gone to great lengths to ensure the quality of this reprint but points out that some imperfections in the original copies may be apparent.

British Library Cataloguing in Publication Data

Developments in petroleum engineering.—1.
1. Oil reservoir engineering
622′.3382 TN871

Visit the Taylor & Francis Web site at
http://www.taylorandfrancis.com

and the CRC Press Web site at
http://www.crcpress.com

PREFACE

Over the last decade or so the advances in petroleum engineering have been described in an ever-increasing deluge of papers prepared for local, national and international meetings. Publication is often erratic, with papers initially being only available to those who attend a particular conference and proving difficult to find at a later date. As a result, it is often difficult to know where to start to get a sensible, and up-to-date, review of a particular subject. This new series of volumes—*Developments in Petroleum Engineering*—is intended to fill this gap and will contain reviews of recent developments. The chapters are written by specialists at a level which summarises the progress, but does not necessarily cover every facet and detail, of a particular subject. Rather, they direct the reader to the most useful of the original sources.

Any assessment of the progress of the exploration and petroleum production divisions in the industry must take into account various factors. The first is that over 60 per cent of the world's energy utilisation comes from oil and gas—which are non-renewable finite resources. The second is that the recoverable reserves for many oil provinces should be capable of extension by proper application of efficient reservoir depletion methods. The third is that the average recovery factor of oil from producing reservoirs does not appear to have greatly increased over the last three decades. The fourth is that although the principles of oil recovery appear better understood, the detailed behaviour of reservoirs is still unpredictable in spite of improvements in data acquisition and interpretative concepts. Finally, that there has been an explosion in computer-based reservoir analyses which combine physical flow principles with detailed geophysical and geological reservoir descriptions which now provide management with the basis for considering alternative development strategies. All the contributions in *Developments in Petroleum Engineering* will consider the progress in the light of these points.

In this first volume we start with the reservoir and its characterisation.

The reservoir is a complex geological area with, initially, only a very few data description points. However, the petroleum engineer does not in general have to have fine geological detail but rather a deep understanding of the trends which cause variations of porosity and permeability, which affect reservoir productivity and development strategy. These trends will guide his decisions in planning reservoir development. This is emphasised in the chapter.

Data from the wellbore are the major sources of information for reservoir delineation, but nowadays the manipulation and analysis of these data have become, in many instances, so specialised that many individuals are dedicated to only one particular task. We have included three aspects: those of well logging, recent hardware developments for well testing and new analysis techniques for the data obtained from well tests. Well logging provides data on the near wellbore description of reservoir lithology, porosity and fluid saturations. Sophisticated instruments are continually appearing as an improved understanding of the near-borehole physics occurs. Well testing is the process whereby the pressure response at the wellbore is observed after a volume of reservoir fluid has been produced. The configuration and physical properties of the reservoir away from the wellbore influence this response. New hardware has continually been brought on to the market, but is not often described, except in trade literature. The analysis of the pressure response is a specialised 'art', and a chapter is devoted to describe its role in management decisions.

Improvements in the numerical methods for reservoir analysis and simulation have occurred along with the new generations of larger and faster computers. Two chapters are devoted to the development of numerical techniques used for reservoir simulation codes. Finite difference schemes are the mainstay of the multiphase multidimensional simulator, and finite element methods are used in some special cases.

Finally, although thermal enhanced oil recovery is the most widely used of the enhanced oil recovery (EOR) methods (some 80 per cent of EOR production is due to thermal techniques) and descriptions of the processes involved abound, the requirements of the physical property database are often overlooked. This chapter attempts to rectify the situation and brings together a collection of suitable sources of information.

The contributors to this volume are experts in their particular field and the editors are extremely grateful to them, especially for their care in the preparation of their chapters and for their patience before the final production.

R. A. DAWE
D. C. WILSON

CONTENTS

LIST OF CONTRIBUTORS

J. R. APPLEYARD

Exploration Consultants Ltd, Highlands Farm, Greys Road, Henley-on-Thames, Oxon RG9 4PS, UK

JOHN S. ARCHER

Head, Petroleum Engineering Section, Department of Mineral Resources Engineering, Royal School of Mines, Imperial College of Science and Technology, London SW7 2BP, UK, and ERC Energy Resource Consultants Ltd, 15 Welbeck Street, London W1M 7PF, UK

I. M. CHESHIRE

Exploration Consultants Ltd, Highlands Farm, Greys Road, Henley-on-Thames, Oxon RG9 4PS, UK

E. DIEMER

Flopetrol Johnston Schlumberger, BP 592, 77005 Melun Cedex, France

A. C. GRINGARTEN

Scientific Software-Intercomp, 1801 California, Denver, Colorado 80202, USA

M. J. PASSMORE

Department of Mineral Resources Engineering, Royal School of Mines, Imperial College of Science and Technology, London SW7 2BP, UK. Present address: *Exploration Consultants Ltd, Highlands Farm, Greys Road, Henley-on-Thames, Oxon RG9 4PS, UK*

JOHN RAE

Theoretical Physics Division, Building 424.4, AERE Harwell, Didcot, Oxon OX11 0RA, UK

P. THREADGOLD

Director, Tesel plc, Houndsmills Road, Basingstoke, Hants RG21 1AJ, UK, and Coombe Cottage, Overton, Nr Port Eynon, Gower, West Glamorgan SA3 1NR, UK

D. C. WILSON

Technical Director, ERC Energy Resource Consultants Ltd, 15 Welbeck Street, London W1M 7PF, UK. Formerly Reader in Petroleum Engineering, Imperial College of Science and Technology, London SW7 2BP, UK

Chapter 1

PETROLEUM ENGINEERING: THE STATE OF THE ART

D. C. WILSON*

ERC Energy Resource Consultants Ltd, London, UK

1. INTRODUCTION

The object of this chapter is to discuss the present state of knowledge in petroleum engineering and the direction of research and development over the next twenty years.

When considering the likely path of future developments it is of course necessary to remember the complex political and economic changes that have occurred over the last decade. Before the first oil price explosion of 1973–74, the world had enjoyed several decades during which the price of hydrocarbon energy had slowly decreased, when measured in real terms. The majority of the oil was produced and transported, at will, by the large major multinational oil companies. This had led to rapid economic growth and an attitude of mind which was not energy-conservation minded. For example, plans were being considered in Saudi Arabia for increasing oil production capacity from about 10 million barrels per day to at least 20 million barrels per day and projections of continued growth of world oil demand were usually presented in terms of increases of between 3 and 6 per cent per annum. Political factors then intervened so much so that oil is now regarded as a national resource, with the host nation having total control over its production and, often, its processing.

Projections in 1985 are therefore very different. Not only because of this nationalism, but also because there is a high level of overcapacity on a world basis, the price of oil is under severe downward pressure and is only

* Formerly Reader in Petroleum Engineering, Imperial College of Science and Technology, London, UK.

being sustained by severe production restrictions within the OPEC countries. Instead of the planned 20 million barrels per day Saudi Arabian production is currently no more than 5 million barrels per day. High cost North Sea production is now the market leader from a price point of view. Indeed it is salutory to remember that total production costs in the Middle East onshore are probably less than US$1 per barrel compared with the US$10 to US$20 per barrel that is more the norm for the most recent developments in the North Sea.

In dollar terms the price of oil peaked in 1979 at approximately US$40 per barrel and has since fallen back to the current levels of less than US$30 per barrel. With the advantage of hindsight it is easy to see that the lead times associated with measures for energy substitution and improvements in the efficient use of energy, combined with the economic slowdown to which high energy prices have contributed, inevitably led to the current situation.

The consequences of these economic and political changes can be clearly seen in the number of enhanced recovery projects that are in operation today in relation to the projections that were being made when the price of oil appeared to be on an ever increasing spiral.

Nevertheless, it cannot be emphasised too much that we are dealing with a finite resource. It is generally agreed that the following figures are a reasonable representation of the world resource base of conventional oil.

(a) An original resource base (including yet to be discovered oil accumulations) of some 5500 to 6500 billion (10^9) barrels of conventional oil. Conventional oil is that which will flow naturally if the applied pressure differential is sufficient and the specific gravity is above 20–25 degrees API.
(b) A worldwide average recovery factor of about 30 per cent, predicted on the basis of the application of the normal primary and secondary recovery techniques.
(c) A postulated ultimate recovery of some 1650 to 1950 billion barrels of conventional oil.
(d) An estimated ultimate recovery of some 1200 billion barrels based on discoveries to date.
(e) A cumulative production of some 570 billion barrels at the end of 1984.
(f) A postulated remaining reserve of 1100 to 1400 billion barrels of which only 630 billion barrels have already been discovered.
(g) A current annual consumption of some 20 billion barrels.

On the basis of the reserves suggested above we can rapidly come to two conclusions. First, a considerable effort is going to be needed in the exploration phase if the yet to be discovered reserves are to be found. Secondly, fertile areas for research and development effort are the extremely large quantities of oil that remain unrecovered when recovery by primary and secondary methods has been completed.

The petroleum engineer is of course primarily concerned with recovery efficiency and it is to a discussion of this that the rest of this chapter is devoted.

2. THE INTEGRATED APPROACH

'Synergy' is the word that has become most popular to describe what may seem to many to be a statement of the obvious—namely, that it is necessary for all the disciplines within exploration and production to cooperate and work together in order that recovery factors may be maximised. In no particular aspect of petroleum engineering is this more necessary than in the case of reservoir simulation. The advent of ever faster computers and the creation of much more sophisticated computer programs to describe the dynamics of reservoir fluid flow is of little avail unless there is a corresponding increase in the interpretative capabilities of both the geophysical and geological disciplines. In fact, the advent of reservoir simulation has not apparently been accompanied by any significant change in recovery factors. There are many engineers who, after the euphoria of the early years of simulation, have now come to the conclusion that the effect of the early attempts at modelling was to reduce recovery factors. This was caused by simplifying the reservoir description to a point that was unrealistic and as a result reducing the required well density for efficient recovery. While there are reservoirs which are sufficiently simple to justify large well spacings it is still probably true that the most potent method of increasing the recovery factor is to drill infill wells.

In recent years, however, the advances in geophysical interpretation with the advent of three-dimensional techniques and the much greater understanding of depositional processes on a reservoir scale have led to much better reservoir descriptions being available on a macro basis.

The current limit of seismic resolution is approximately 10 to 20 metres and it will require major advances in technique before this can be improved. Similarly, the study of any outcrop is a revelation to the geologically uninitiated and illustrates the complexity of the geological input from the point of view of reservoir description. Nevertheless our basic understanding of geophysical and geological processes has advanced to the point where the advances in reservoir simulation described in later chapters of this book can now be used to their full advantage.

The petroleum engineering contributions to advances in reservoir description have centred (other than simulation) on advances in petrophysical interpretation, including *in-situ* point pressure measurements, and in pressure analysis techniques. It is important that petrophysical information and analysis be combined with core-measured values of porosity, mineralogical content of the matrix and the general description of sedimentary structures as the problems are, in general, ill posed with multiple solutions and only the use of all available information channels will give satisfactory results. Very similar comments can be made about pressure analysis techniques where it is vital to have a geological model in mind when trying to interpret a specific pressure response. By way of example an upward slope on a Horner plot can be caused by many factors but two possible interpretations would be either declining permeabilities away from the well bore or else an actual formation boundary. The decisions taken might be very different depending on which model represented the preferred explanation for the pressure behaviour.

Nowhere more than offshore, for instance the North Sea, can it be said that the integrated approach is vital. It is instructive to consider why this is so and to see what lessons for future developments can be deduced. Several points of difference occur between the major offshore developments and their counterparts onshore. Perhaps the most obvious difference is the level of appraisal that is undertaken before major investments are made. In most North Sea fields currently on production the development decision was taken after no more than five exploration and appraisal wells had been drilled; that point was reached in the case of Mobil's Beryl field after only two wells had been drilled. In a land development, on the other hand, providing government regulations do not stop piecemeal development, there is an almost overwhelming argument for a stepwise approach.

Why is this? The simple answer is that the stepwise approach is the correct one under all circumstances if economic considerations are eliminated. Production performance under primary depletion is the most revealing indicator of reservoir continuity and in all but the simplest of reservoirs is absolutely necessary if a secondary recovery scheme is to be designed to the maximum level of efficiency. It is here that the multiple pressure recorder or, as it is commonly known, the repeat formation tester, comes into its own. However for offshore developments, where a platform production system is needed, the huge 'front-end loaded' investments are such that the interest rates alone are greater than many countries' gross national products. The platform and its production facilities must be planned and executed before there is much, if any, production history and the production rates must be such as to ensure a reasonable rate of return on investment. There is, therefore, a second difference between previous developments and that of recent ones such as the North Sea—that is the rate of depletion. The North Sea is unique in that giant oilfields are being produced at a peak rate of 10 to 15 per cent of ultimate recovery per annum. This compares with an average rate in the Middle East of less than 2 per cent per annum and in the case of some of the largest fields of less than 1 per cent per annum. Time is therefore not on the side of the petroleum engineer and it is perhaps just as well that the tools at his disposal have been improved over the last decade.

3. ENHANCED OIL RECOVERY

As stated in Section 1 the worldwide ultimate recovery factor, assuming primary and secondary recovery only, appears to be no more than 30 per cent. Therefore, there is an immediate incentive to look for recovery methods which would have an appreciable effect on this.

Simply increasing the recovery factor to 45 per cent, at first sight perhaps an apparently not too ambitious target, would have the effect of increasing the known world remaining reserves by a factor of nearly two. The same point can be made by considering that increasing the recovery of the Forties field in the British Sector of the North Sea from 50 to 60 per cent would be equivalent to finding another eight fields of the size (about 50 million barrels of reserves) currently being considered as marginal candidates for development.

Before discussing the state-of-the-art position as regards enhanced oil recovery it is perhaps as well to formalise the definition of some terms.

Primary recovery is that which occurs due to the natural energy of the reservoir system and is based on the expansive capabilities of the reservoir rock and the reservoir fluid.

Secondary recovery is that which occurs when energy in the form of water or gas is artificially introduced into the reservoir in order to maintain pressure and to move the oil to the production wells.

Tertiary or *enhanced* recovery (EOR) consists of applying further recovery techniques which are devised by the petroleum engineer and which do not mimic naturally occurring processes in the reservoir. In this sense conventional gas injection is not an enhanced recovery process as it merely mimics solution gas drive or gas cap drive. Similarly, water injection is not an enhanced recovery process as it merely mimics aquifer drive.

EOR processes can be divided into three main categories:

(a) Thermal processes consisting of steam flooding, cyclic steam injection, hot water injection and *in-situ* combustion.
(b) Chemical processes including surfactant combined with polymer injection, polymer flooding and caustic flooding.
(c) Miscible displacement processes including miscible hydrocarbon injection, carbon dioxide injection and nitrogen injection.

Steam flooding applications are by far the most numerous and, with certain notable exceptions to be mentioned later, the only wholly proven method of EOR from an economic point of view. The principle, if not the execution, is extremely simple and very similar to pattern water flooding except that it is the application of heat energy rather than pressure energy that is the main contributor to improving the recovery. The energy source is the latent heat contained in the steam and the main effects are thermal expansion of the oil, combined with large reductions in viscosity, causing the oil to flow much more easily. Secondary effects include reduction of residual oil saturations by distillation of the lighter components and changes in relative permeabilities. Most developments are based on very close well spacing and, as such, are usually restricted to areas where such spacings are economic or already drilled. In California, for example, 5 acre spacing or less is the norm. Further limiting factors are depth and reservoir thickness. The depth limitation is caused by the critical pressure of steam which at 218 bar (3200 psia) restricts reservoir depths to about 2000 m (6000 ft) as an absolute maximum while the thickness limitation is related to loss rates of heat to the over- and under-burden which makes thin reservoirs (say less than about 5 m) very inefficient. Hot water injection can be used but has, thus far, been beset with problems of premature breakthrough caused by lack of mobility control.

In situ combustion consists of heating the reservoir fluid by means of combustion gases which pass ahead of a burning front. The primary requirement is, again, close spacing as a flow path for the gas must be established prior to ignition in order that oxygen can be continually supplied to the combustion zone. There are several variations on the process. Wet and partially quenched combustions seem to offer the most effective variations as they involve *in situ* generation of considerable quantities of steam which itself aids the recovery process. The products of combustion include considerable quantities of acid gases resulting in difficult corrosion problems.

The chemical processes have shown significant potential in the laboratory but as yet have proved disappointing in field trials. It is evident that although the potential of surfactants to reduce the interfacial tension of an oil–water interface to near zero values can be easily demonstrated, the problems associated with applications under reservoir conditions have yet to be overcome. Furthermore, the problem of adsorption of surfactants on the reservoir rock is an important consideration as is the effect of saline conditions combined with temperature and pressure on the stability of both the surfactant and the chase polymer. In the case of polymer, degradation due to shear, bacteria and free radicals are also major problems.

One further difficulty which is often ignored is the question as to the exact disposition of residual oil. Surfactants can only work to remove residual oil if the distances are very small so that if the oil is in larger clumps the surfactant can only 'nibble' round the edges. Detailed descriptions of swept and unswept areas of the reservoir are absolutely essential. The effects of heterogeneities, especially those within supposedly uniform sands, are, in reality, still totally unknown. A major research effort, both theoretically and with pilot field trials, is absolutely essential.

Caustic flooding is simply a method of changing the wettability characteristics of the reservoir and of generating surfactants *in situ*. It has the advantage of being inexpensive compared to surfactant injection but it is not very obvious at this stage of development whether worthwhile improvements in recovery will be achieved.

Miscible displacement consists of eliminating all interfacial tension forces between the displacing and displaced fluids so that, in principle, there is no residual saturation of the displaced fluid within the contacted area. In principle, the displacing fluid could be any of LPG, condensed hydrocarbon gases, carbon dioxide, nitrogen, exhaust gases or even certain alcohols. To date there have only been field trials based on gas or gas liquids injection. LPG has been used most often and, in certain reservoirs of high relief with good gravity segregation, it has been very successful with recovery factors in excess of 80 per cent being predicted. In these cases a gravity stable sequence of oil, LPG and dry gas banks are formed in the reservoir and move downwards towards the producing intervals. This process is known as first contact miscibility.

All the other injection gases are initially immiscible unless the temperature and pressure are very high and the miscible bank is formed by a complex process of stripping of the light ends of the reservoir oil to form a bank of miscible fluid by a multi-contact process.

Carbon dioxide is seen as having a high potential in certain parts of the USA where there are naturally occurring accumulations. It must be remembered, however, that the gas is extremely corrosive as carbonic acid which is inevitably formed in large quantities and special precautions must be taken to limit corrosion. Asphaltine precipitation is a possible difficulty, although this may, ultimately, only be a laboratory problem.

Nitrogen is being considered seriously for miscible displacement in some of the North Sea fields but no field trial has yet taken place. Density differences may cause gravity override and the low viscosity may cause viscous instabilities, so limiting the potential applications. New forms of surfactants to act with nitrogen as foam mobility controls are actively under consideration. The advantage of nitrogen is that it can be obtained cryogenically from the atmosphere, although the energy used to produce the gas at the required pressure may be less than the energy equivalent obtained by the increased oil production.

From the above discussion it is seen that, although there are many enhanced recovery techniques being applied, the methods (with the exception of steam flooding) can only be described as interesting; they are barely past the laboratory or field trial stage.

4. THE FUTURE

It would be very unwise to predict that the future advances in petroleum engineering will be other than slow and painful. The major problems of increasing recovery have yet to be solved and many of the techniques described in the previous section have little demonstrated economic viability. It is well to remember that about 60 per cent of the world's known remaining oil reserves are contained in about 25 fields and it is what happens to the recovery factor in these fields that will have most impact from a global point of view.

The continuing necessity for improvements in reservoir description cannot be overemphasised and it is perhaps in this area that the main hopes lie. If we cannot properly evaluate the likely response of a reservoir to the simple process of gas or water injection there is little chance that we can predict the response to the more complicated methods now being proposed.

One barrier to continued advancement is of course the huge price disparity between advanced recovery techniques and that of waterflood recovery from the large onshore fields of the Middle East. It requires an enlightened attitude to the long term and the eventual scarcity value of oil to justify large expenditures on research and development at present. Politicians are notorious for their short-term view of the future and research which is unlikely to bear significant fruit in the next decade will be difficult to fund.

There will of course be many gradual improvements in equipment and operating procedures. The development of the downhole multiple pressure recorder is an example; the quartz pressure gauge giving direct surface readout is another. Other chapters in this book record in detail some of these types of equipment developments and methods of analysis. The improvements in deviated drilling have enabled offshore platforms to be used with great effect and for larger radii of depletion, but with the penalty of larger well spacings than used onshore. Developments in sub-sea equipment and tie-back facilities may overcome some of these deficiencies. Unfortunately, the limitations of space and weight on platforms may mean that improvements in facilities on older (producing) fields may not be rapidly implemented. However, more compact and efficient separator and primary processing equipment is currently under development, especially for computer-controlled automated and unmanned stations.

Drilling techniques are undergoing many changes, especially with the introduction of new bit design, new low toxicity, oil-based drilling fluid systems and well bore trajectory control and telemetry techniques. Horizontal drilling methods will enable heavier oils to be produced at possibly more economic rates. Concern for improving the borehole conditions for general stability and for logging are continually being reported: this is especially needed for the deviated and horizontal wells. Reliable production logging methods are still required for these wells though.

As the commercial value of gas increases, optimisation of the processing facilities to reduce the waste of the lighter hydrocarbons will become more sophisticated especially with the increase in the numbers of condensate reservoirs coming onstream. Already, with the improved seismic and other geological methods, and deeper exploration drilling, condensate fields are becoming important prospects for hydrocarbon supplies. However, the choice of development plan is still totally uncertain and much careful analysis of such effects as compositional mass transfer within the porous matrix of the reservoir, processing methods and pipeline transport must be made before the optimal production method can be put forward with confidence.

There are still great improvements needed in many other parts of the oil winning processes. The spectres of corrosion and bacteria control are high on everybody's list, but better metering, loss control (due to evaporation, poor separator design or just leaks), sampling, safety, especially when drilling, blow-out prevention and kick detection and even batteries that can withstand high temperatures are all in need of attention. Improvements here will be gradual, but with occasional inputs from the regulating authorities. Environmental concern, now vocal in many parts of the world, will have to be taken seriously and appropriate action taken. Many of the objections are ill founded but some of the points are justified and correction is essential.

This chapter has concentrated wholly, thus far, on oil resources as it is in that area that the main scope for change exists. It would be wrong, however, to conclude without discussing the role of natural gas in the future. Recoveries from reservoirs of even moderate permeability have been high, with recovery factors in excess of 80 per cent being commonplace. The scope for increase is thus severely limited. Where advances will be made and where extra resources will become available will be in the development of the large gas accumulations where the permeability can be measured in microdarcies rather than millidarcies. Cost is again the main limiting factor but advances in fracturing techniques will play a major part in unlocking this large resource.

Furthermore, though it is debatable whether this is a petroleum engineering topic being largely one based on advances in mechanical and civil engineering, it is inevitable that the exploration and development end of the oil industry will continue to probe the frontier areas. By the end of the century oil production from water depths in excess of 500 m will be relatively commonplace and production of significant quantities of gas from this water depth will have commenced.

Finally, as the difficulty of securing further supplies of the resource increases, the number of professionals required in the industry will increase. It is therefore essential that those individuals already in the industry make sure that the future is portrayed as exciting and dynamic in order that the new entrants are in sufficient numbers and of sufficient quality that we may look forward with confidence to the time when the easy-to-recover hydrocarbons are all but exhausted.

Chapter 2

RESERVOIR VOLUMETRICS AND RECOVERY FACTORS

JOHN S. ARCHER

Department of Mineral Resources Engineering, Royal School of Mines, Imperial College of Science and Technology, and ERC Energy Resource Consultants Ltd, London, UK

1. INTRODUCTION

This chapter presents a brief review of reservoir volumetrics and recovery factors and some of the methods by which they are currently obtained. The trend in Europe has been towards a probabilistic representation of hydrocarbon in place and recoverable reserves. To a large extent this has resulted from the necessity of assessing risk in the capital funding of large offshore developments. The uncertainty in knowledge of reservoir characteristics at the time when development decisions are made and the relatively long time (several years) before first production, has influenced a more multidisciplinary approach to the estimation of reserves. It is probably true to say that the offshore petroleum engineering industry has its frontier in terms of state-of-the-art technology in North Sea reservoir development. The decisions to finance development are based in general on data from relatively few exploration and appraisal wells and this constraint has resulted in an attempt to maximise data gathering and interpretation. The greatest degree of reservoir description integration has taken place between specialist geologists and reservoir engineers in the area of geological modelling.

Conceptual geological models are being used as a basis for reservoir simulation models in sensitivity studies of reservoir development options. Reservoir simulation has advanced to a degree where the capability to describe physical processes and flow mechanisms in terms of mathematical relationships and to solve the interactive non-linear differential equations which result, has outpaced the capability to describe the reservoir environment. The application of reservoir simulation studies is essentially one of assessing the sensitivity of given uncertain parameters to a preferred development plan in the pre-production phase. Once recovery with a given scheme has started, the reservoir models evolve as operating tools, but are updated to ensure that observed performance is respected in the model. This aspect of 'history matching' is an area of increasing co-operation between development geologists and reservoir engineering specialists and requires consideration of expected and demonstrated flow paths and flow barriers in the reservoir system.

In current North Sea petroleum development activity the interest in hydrocarbon in place obviously provides the first basis for estimation of recoverable reserves and commitment to development. The in-place estimate also provides the most widely used basis for equity share between partners in a given reservoir. It is extremely rare to find equity agreements (in contrast to proposals!) with any basis involving moveable hydrocarbon. The particular interest in recoverable hydrocarbons arises naturally from the economic evaluation of a proposed development. With relatively conventional recovery processes such as powered water injection the representation of a recoverable reserve probability distribution tends to be tied to that mechanism and other primary or enhanced processes are excluded.

2. PROBABILISTIC REPRESENTATION OF HYDROCARBONS IN PLACE

The currently employed terminology between operating companies, financial institutions and independent consultants has evolved in North Sea development as an association of words such as 'Proven' with a probabilistic expectation that a value will not be less than the defined amount. The following terminology was used[1] in the well-publicised 1982 privatisation sale of the interests of Britoil plc:

> *Proven:* in relation to quantities of petroleum, means the amount thereof which geophysical, geological and engineering data indicate to be in place or recoverable (as the case may be) to a high degree of certainty. For the purposes of this definition, there is a 90 per cent chance that the actual quantity will be more than the amount estimated as Proven and a 10 per cent chance that it will be less.

Probable: in relation to quantities of petroleum, means the amount thereof which geophysical, geological and engineering data indicate to be in place or recoverable (as the case may be) but with a greater element of risk than in the case of Proven. For the purposes of this definition, there is a 50 per cent chance that the actual quantity will be more than the amount estimated as Proven + Probable and a 50 per cent chance that it will be less.

Possible: in relation to quantities of petroleum, means the amount thereof which geophysical, geological and engineering data indicate may be in place or recoverable (as the case may be) but to which a substantial element of risk must be attached. For the purposes of this definition, there is a 10 per cent chance that the actual quantity will be more than the amount estimated as Proven + Probable + Possible and a 90 per cent chance that it will be less.

These data may be represented graphically as shown in Fig. 1 as the cumulative probability that the value shown will be not less than the indicated value of hydrocarbon in place or recoverable reserve.

The probabilistic representation is recognised as helping to define uncertainty. Other definitions of the words 'proven', etc., are being suggested by a working committee of the Society of Petroleum Engineers in the USA and by a group set up to report to the 1983 World Petroleum Congress.[2] None of the definitions have found universal acceptance.[44] Part of the problem is that the assessment of risk and probability, although made from technical considerations, is essentially subjective. The trend in North Sea reservoir evaluation is to agree probability distributions for parameters with uncertainty in multidisciplinary committee. That is to say, specialists are 'forced' to assess and justify their representation of reservoir description as a group of geologists, geophysicists, petrophysicists and reservoir engineers. The simplest representation of the reservoir condition (temperature and pressure at datum) hydrocarbon in place is the hydrocarbon-filled pore volume V_{hc}. In units of reservoir cubic metres this is

$$V_{hc} = Ah\phi S_{hc}$$

where A is the area (m^2) representing the reservoir, h is the average net thickness (m) in the reservoir, ϕ is the fraction of the bulk rock volume representing the pore space available to reservoir fluids (oil, water, gas) and S_{hc} is the fraction of pore space in the defined reservoir which is filled with hydrocarbon.

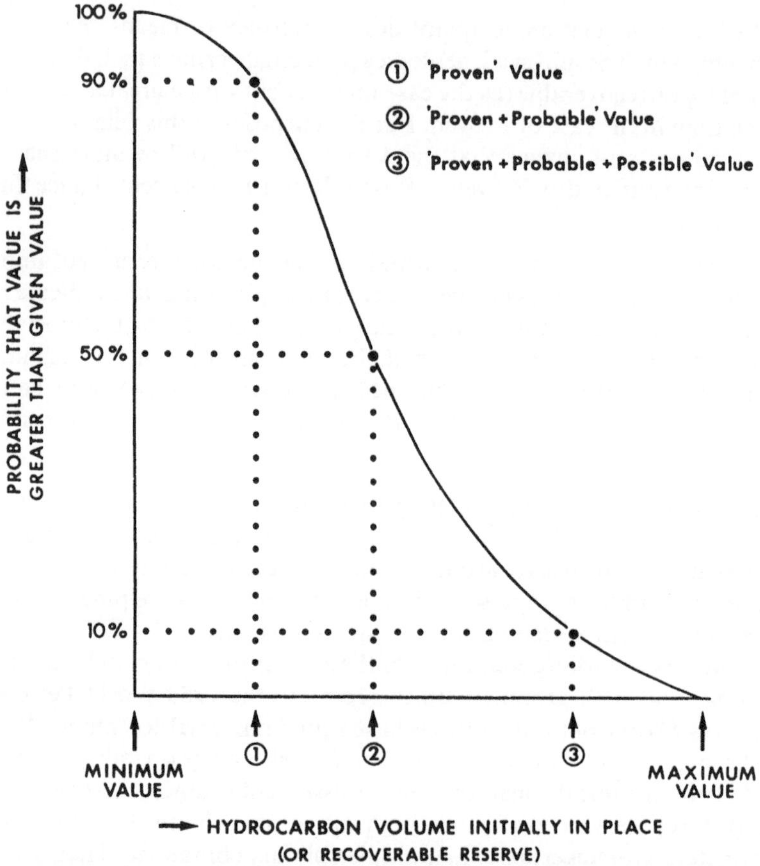

FIG. 1. Probabilistic representation of original in place hydrocarbon volume and recoverable reserves. Reproduced by permission of ERC Energy Resource Consultants Ltd.

When single 'best' estimates of each parameter are made the hydrocarbon in-place value obtained is known as a deterministic estimate. It is clear that where areal and vertical variation of the reservoir properties occur the calculation can be subdivided to reflect zonation. For a reservoir with subzones $j = 1$ to $j = k$, the total in-place estimate of V_{hc} is:

$$V_{hc} = \sum_{j=1}^{j=k} (A_j h_j \phi_j S_{hc_j})$$

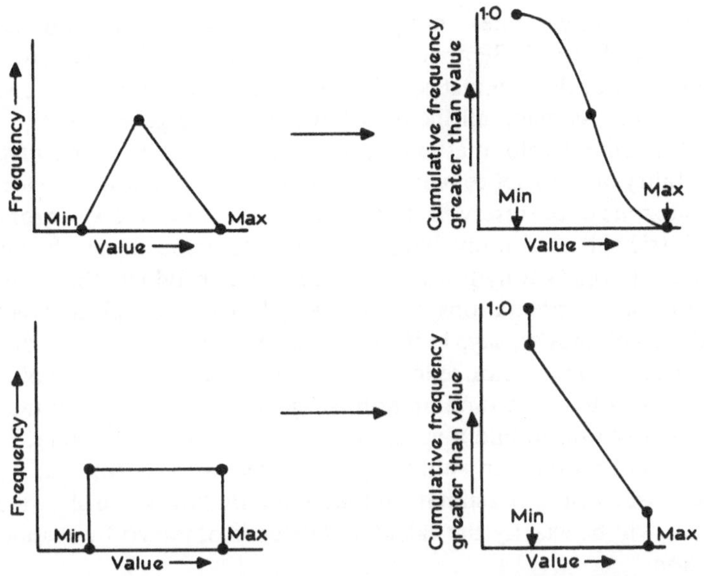

FIG. 2. Distributions of reservoir properties: actual and cumulative.

It is further clear that the representation of each parameter of the V_{hc} calculation is subject to an uncertainty which may decrease as more reservoir data are acquired.[3,4]

One of the more successful techniques in attempting to quantify the probabilistic distribution of hydrocarbon in place or reserves is through application of Monte Carlo analysis. The essence of the method is an unbiased sampling of each defined distribution of the independent components of the volumetric equation and the calculation of chance-related values. Each independent variable in a region of the reservoir must therefore be defined probabilistically. With relatively little well control such distributions often follow simple shapes with triangular and rectangular distributions finding most favour as shown in Fig. 2. Normal distributions, log-normal distributions and skewed distributions find their place in particular reservoir representations.

It is important to recognise that the distributions have maximum and minimum values in reservoirs and do not go to plus or minus infinity. The triangular distribution tends to reflect the conceptual 'most likely' plus maximum and minimum values and is very frequently applied to areal and thickness representations. The porosity and saturation terms in a hydrocarbon zone should perhaps be considered as a product rather than as independent variables since porosity versus saturation relationships are frequently observed. Before considering the source and uncertainty in the components of the volumetric calculation the Monte Carlo method will be summarised assuming all the distributions can be generated.

The Monte Carlo method requires a frequency of occurrence or probability of occurrence representation for each independent variable for each subzone of the reservoir. For a reservoir divided into k subzones and a volumetric calculation involving subzone area, subzone net thickness and subzone porosity × hydrocarbon saturation product the number of distribution representations is $3k$. If the hydrocarbon in place is required in standard volumes at, say, 15 °C and 1 bar pressure, then the hydrocarbon formation volume factor representation increases the number of required distributions to $4k$. A random number generator is utilised to interrogate each distribution to find a value associated with its probability. For each independent variable in subzone j associated with its randomly selected probabilistic value, an estimate of the chance that the value of hydrocarbon in place will be exactly the calculated value is obtained by evaluating the equation

$$V_{\mathrm{hc}_j}(p_1 p_2 p_3) = A_j(p_1) h_j(p_2) \phi S_{\mathrm{hc}_j}(p_3)$$

where (p) indicates the probability that the value will be exactly the calculated value. The calculation is performed a large number of times, generally in excess of one thousand, in order that the distribution associated with each independent variable has been fully sampled. From the number of occurrences of particular values the cumulative probability distribution is calculated. The combination of subzone values into a total reservoir representation is performed by selecting a probability level for each subzone and summing the indicated value. If, for example, in subzones 1, 2, 3 at the 40 per cent cumulative probability level the values of oil in place (in millions of reservoir cubic metres) have been calculated as 50, 94 and 106, then the cumulative total reservoir representation at the 40 per cent level the value would be shown as 250 million reservoir cubic metres.

3. UNCERTAINTY IN RESERVOIR DESCRIPTION

The sampling of any reservoir in order to determine a continuous representation of the environment is unreliable. The actual volume of sample possible is insignificant, as shown in Fig. 3, in comparison to the unsampled reservoir volume and is never statistically random. In order to represent the

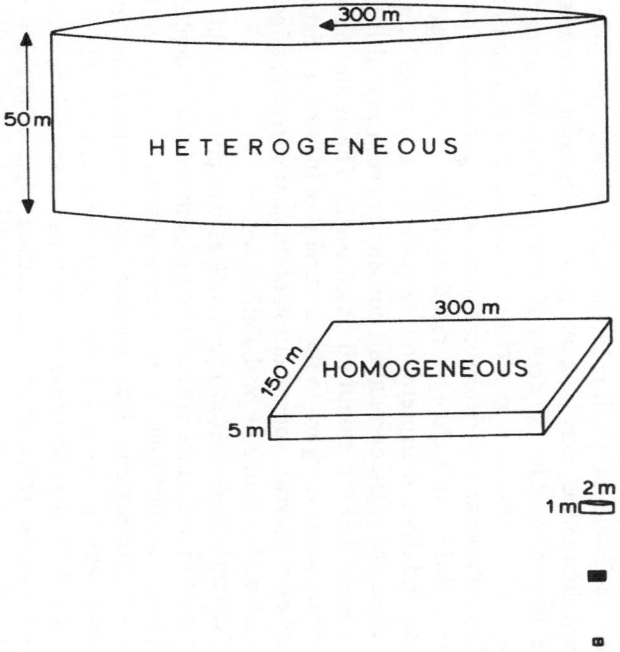

	Bulk volume (m^3)	Relative volume
Well test	14×10^6	10^{14}
Reservoir model grid cell	$0{\cdot}2 \times 10^6$	10^{12}
Wireline log interval	12	10^8
Core plug	2×10^{-5}	200
Geological thin section	1×10^{-7}	1

FIG. 3. Relative scale of raw and applied reservoir description data.

hydrocarbon pore volume of a reservoir it is necessary to define the boundaries of the system, the volume of the pore space and the fraction of the pore space occupied by petroleum fluids. At any well a vertical (or at least convertible to a vertical) record of rock character may be obtained by coring and wireline logging, with information augmented by drilling performance, mud logging and cuttings. The information recorded by the sensors in wireline logs is also dependent on tool and wellbore condition and signals are converted to petrophysical properties by mathematical relationships which are assumed to be applicable. Some of these data will now be considered.

3.1. Porosity and Saturation in a Well

The interpretation of wireline logs to obtain porosity and saturation values depends on the recognition of wellbore bulk lithology and the (sometimes) subtle influence of associated minerals. In North Sea reservoirs the effects of heavy minerals, micas and the surface area effects of detrital and authigenic clays complicate analysis. The identification of lithological boundaries can sometimes be problematical and is associated with source and detector spacing in some tools as well as logging speed and hole condition. The representation of log-interpreted porosity and saturation from a number of alternative model calculations is frequently made at 0·25 m intervals. These data still reflect averaging techniques over larger intervals even though they may look like positionally exact information. It is also necessary to consider the depth of sampling into the formation by wireline log tools where, as a generality, depth of investigation may be considered as proportional to detector–source spacing. The penetration of mud filtrate together with changes resulting from its properties in the reservoir can influence interpretation of porosity and saturation. The problem is increased when correlating porosity between wells drilled with water-based muds and those with oil-based muds. The author exhibits a degree of scepticism with regard to the 'accuracy' of log-derived data. This does not, however, prevent representation of the characteristics of porosity and saturation with a range of probabilistic values.

Core-derived information from a given well at least provides point values for porosity assuming that the core can be properly related to depth in the reservoir. There are, however, at least as many problems in obtaining representative information. For routine core analysis information (porosity, permeability, grain density and 'as received' saturations) the problems are considerably less severe than those in special core analysis. Routine core analysis methods involve cutting plugs about 2·5–3·7 cm diameter from the whole core. Often the plugs are cut at 90° to the whole core or parallel to bedding planes. Plugs may be selected at approximately 25 cm intervals along the whole core. End faces are trimmed off by sawing with the intention of removing those sections invaded by drilling mud. The resultant plugs may be from 2·5 to 7·5 cm long depending on test requirements.

The cutting of plugs may require attention to bit coolant and wash fluids, as kerosene can cause asphaltic plugging with some crude oils and fresh water or low sodium salinity brines can cause porosity/permeability modification in rocks containing certain clay minerals. The author has observed porosity–permeability increase through solution during plug cutting of southern North Sea sandstones containing high proportions of halite. Routine core analysis plugs are frequently cleaned of hydrocarbons and pore waters by reflux with organic solvents such as toluene, hexane, heptanes and methanol. It is important that the pore space is unaltered by these processes, which may be difficult with sensitive clay materials. Where the pore water has been of high salinity, salt crystallisation may occur during core plug drying and provide an erroneous contribution to bulk volume. The drying of core plugs after organic cleaning may be performed in a nitrogen atmosphere at around 500 °C. Water-sensitive clays require less severe, humidity controlled, drying otherwise their morphology and surface area may change.

Although permeability is not a consideration in the hydrocarbon in-place calculation it is of course of major importance in recovery calculations. The permeability measurements in routine core analysis have been particularly influenced by the handling processes prior to measurement of the property. The drying process can collapse certain authigenic clays in the pore space and critical point drying techniques have been reported which demonstrate the effect.[5,43] Both porosity and permeability measurements on core plugs are influenced by the choice of measurement technique.[6] For porosity the measurement methods include Boyle's law techniques, summation of fluids and use of the helium pycnometer. The latter technique has the greatest precision and is preferred.[7] Permeability measurements in routine analysis involve measurement of the pressure gradient across the plug as a result of helium or nitrogen flow through it. With high-permeability reservoir rocks the accurate measurement of low differential pressures becomes important. The correction of permeability for Klinkenberg slippage and for inertial or turbulent flow effects is not usually considered as part of a routine analysis. If Klinkenberg correction is requested from a commercial laboratory it is necessary to establish that calibrations with liquid measurements are conducted, as 'empirical chart' corrections are proving to be quite unreliable. In routine core analysis neither porosity nor permeability measurements are conducted at net overburden pressure, and corrections based on pore volume compressibility or of an empirical nature are required before comparison with *in-situ* measurements are made.

The selection of core plug samples from whole core is a matter for proper concern. Many engineers and geologists provide no guidance to commercial laboratories for plug selection. Ideally, a selected plug should also provide a sample for geological, mineralogical and sedimentological analysis. The location of the plug should, as far as possible, provide a 'net sand' location as the data are used in comparison with net porosity from log interpretation.

The accuracy of some petrophysical measurements has recently been reported in a comparative study by Hook.[8] He found that in a statistically significant set of samples, measured by routine core analysis methods for porosity by four different laboratories, there was a deviation around a mean value of 11 per cent of $-0{\cdot}53$ to $+0{\cdot}41$. This alone can provide a 10 per cent difference in estimates of hydrocarbon in place. It is usually believed that routine core analysis techniques can represent porosity to within 0·1 porosity per cent units of a 'true' value.

Saturation representation in a reservoir derives almost entirely from log interpretation and requires particular knowledge of formation water and mud filtrate properties at reservoir conditions in addition to porosity characterisation. Since many sandstones can be considered to have at least some element of 'shaliness' the characteristic of pore surface area and the influence of clay minerals is a necessary part of interpretation. The state-of-the-art at present does not adequately allow such characterisation. Clay properties for V_{clay} are often assumed related to detrital clay or shales even when authigenic clays are known to be significant.[9] Surface area representation such as is required for the Waxman–Smit method and its derivatives[10] is still only obtainable from core chemistry methods. Relationships between log interpreted saturations and those predicted for capillary–gravity equilibrium from core plug measurements are also beset with uncertainty. Minimum values of connate water saturation may sometimes be derived directly from cores drilled with oil-based muds or lease crudes and such data can be correlated with log interpretation and capillary pressure indications. The basis for a capillary pressure application is that at some wetting phase saturation (S_w) the value of capillary pressure (P_c) in a given pore size (r) system is controlled at a given saturation by the interfacial tension (σ) of the fluid–rock system and the wetting character of the fluids as indicated by the contact angle θ, i.e.

$$P_{c(sw)} = \frac{2\sigma \cos\theta}{r}$$

for different fluid systems in the same rock (r = constant)

$$\frac{1}{r} = \left.\frac{P_{c(sw)}}{2\sigma\cos\theta'}\right]_{\text{Fluid system 1}} = \left.\frac{P_{c(sw)}}{2\sigma\cos\theta}\right]_{\text{Fluid system 2}}$$

A laboratory-derived drainage capillary pressure curve for an air–brine system in which σ is well established and for which water is the strongly wetting phase ($\theta \simeq 0°$; $\cos\theta \simeq 1{\cdot}0$) can therefore be used to generate, for example, the equivalent oil reservoir condition capillary pressure at the same wetting phase saturation, i.e.

$$P_{c(sw)\ \text{oil-water, res}} = P_{c(sw)\ \text{air-water, lab}} \frac{(\sigma\cos\theta)_{\text{oil-water, res}}}{(\sigma\cos\theta)_{\text{air-water, lab}}}$$

By defining the position in the reservoir where the pressure in the oil phase is equal to the pressure in the water phase (i.e. $P_c = 0$) as the free water level (FWL) and using this as a depth subsea above which elevations are defined by H, we can write a gravity equilibrium expression in consistent units in terms of the density difference of the fluids at reservoir conditions as

$$P_0 - P_w = P_c = gH(\rho_w - \rho_0)$$

It therefore follows that, for a given reservoir rock of the same pore size distribution, the position in the reservoir at which a particular saturation S_w will occur is given by

$$H_{(sw)} = \frac{P_{c(sw)\ \text{air-water, lab}}}{g(\rho_w - \rho_0)} \frac{(\sigma\cos\theta)_{\text{oil-water, res}}}{(\sigma\cos\theta)_{\text{air-water, lab}}}$$

A particular value of $H_{(sw)}$ that is of interest and which can be inferred from laboratory experiments is that of the position of the equilibrium observed oil–water contact as some elevation above the free water level as shown in Fig. 4. A capillary pressure known as the threshold capillary pressure (P_{c_t}) is needed to displace any non-wetting fluid from the largest available pore in the pore size distribution. This gives rise to an equivalent region of 100 per cent water saturation above the $P_c = 0$ position. In wells, the observed oil–water contact is obtained from logs and the free water level is obtained

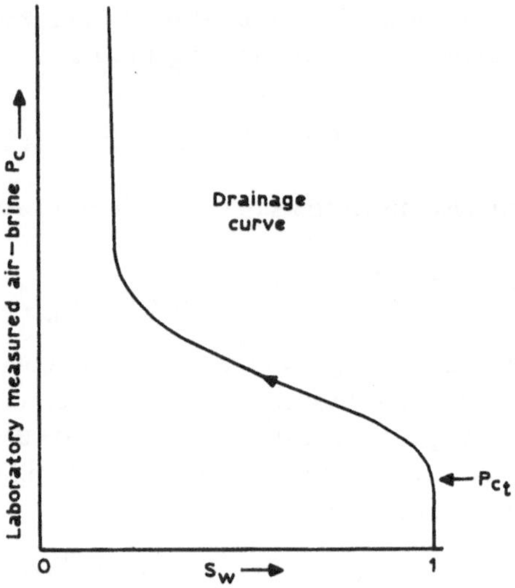

FIG. 4. Laboratory drainage capillary pressure curve.

from laboratory P_{c_t} measurement or from intersection of fluid gradients using, for example, RFT data.

When more than one rock type (i.e. pore size distribution) occurs in a reservoir it should be expected that superposed capillary pressure curves could define the equilibrium initial saturation distribution, as shown in Figs 5(a) and (b). In fact, the observation of a number of different irreducible water saturations above the transition zone is also a good primary indicator of rock character variations and the degree of subzonation required to define the reservoir.

In case the technique described sounds too simple it is also worth remembering that we have in general no adequate knowledge of σ or θ for the reservoir fluid, reservoir condition system, and they are usually guessed! The default value assumed by many commercial core analysis laboratories assumes water-wet rock gives $\sigma \cos \theta = 26\,\text{dyne/cm}$. The inability to measure this value results partly from the uncertainty in knowledge of wettability change during coring and core recovery and the difficulty in measuring equilibrium contact angles on reservoir rock surfaces. The residual saturation inferred from mercury displacing air capillary pressure tests is rarely the same as the true wetting phase irreducible saturation and the mercury tests should be used mainly for pore size distribution measurement (see Fig. 6).

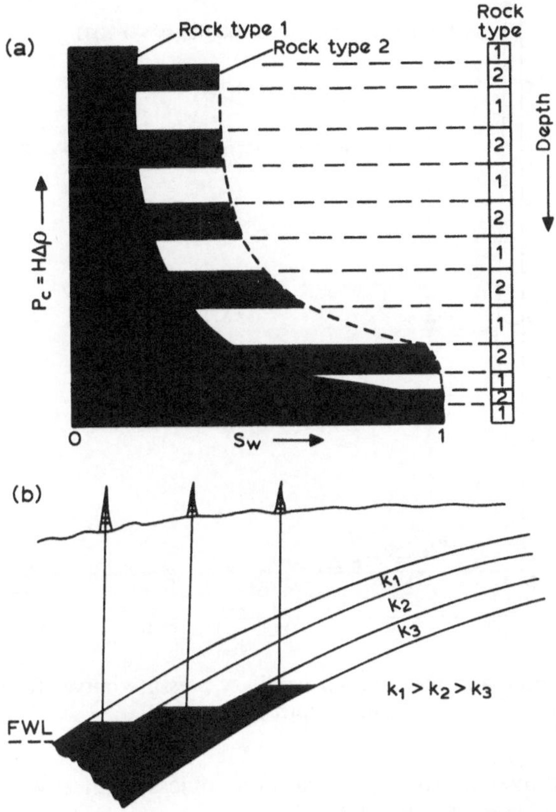

FIG. 5. (a) Saturation distribution in a two rock type reservoir; (b) observed oil–water contacts in a reservoir with different rock properties.

3.2. Net Thickness in a Well

In a well the net thickness determination is based on true vertical depth (TVD) or true vertical thickness (TVT) of the reservoir interval as calculated from wireline logs, and from local definition of net reservoir quality. The geological description of primary lithology, for example sand or shale, is generally insufficient in defining the petrophysical net:gross thickness ratio. The petrophysical characterisation of net quality in sandstone reservoirs generally incorporates an exclusion of rock with V_{clay} greater than, say, 40 per cent or water saturation greater than, say, 60 per cent. In addition, in a given 'rock type' zone an empirical relationship between porosity and permeability may lead to the identification of

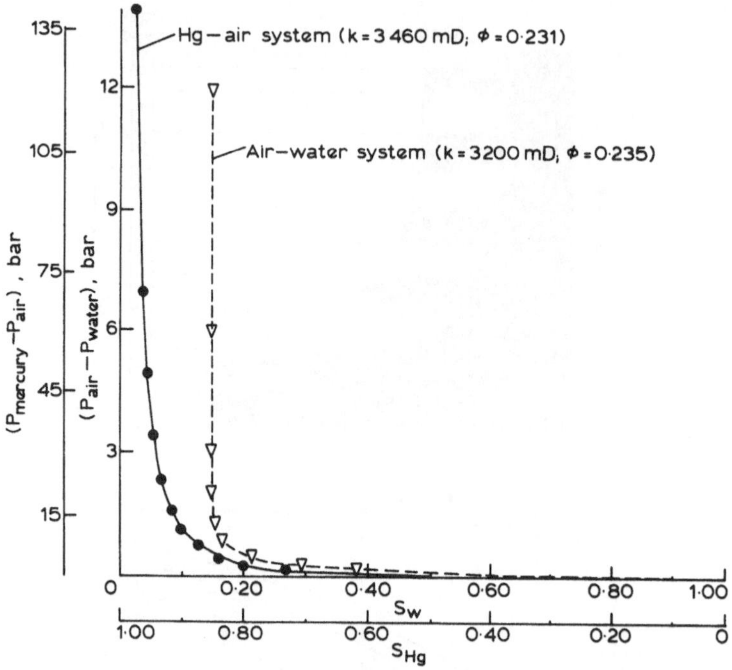

FIG. 6. Air–brine and mercury–air capillary pressure curves from plugs at the same depth of sample.

porosities equivalent to a permeability of less than say 1 mD as non-productive. This particular definition should recognise the nature of the mobile fluid since gas can move with much greater ease than oil in 1 mD rock, and at an oil–water contact with large areal extent a large volume of water can move vertically across a 1 mD permeability interface.

The preceding discussion on porosity, permeability and saturation is applicable to the quantitative characterisation of net thickness at a well. A particular aspect of thickness characterisation which leads to an element of uncertainty involves the geometry of the reservoir sediments. It is quite possible for the thickness of a sand or shale to increase or decrease (or to disappear altogether) at short distances from the well. It is necessary to associate with a given reservoir rock a conceptual understanding of its likely depositional geometry. This can be inferred from detailed sedimentological study of recovered core and is a necessary integrated aspect of reservoir evaluation. In order to illustrate this further, Fig. 7 shows a 0·5 m shale interpretation from a gamma-ray log.[11,12]

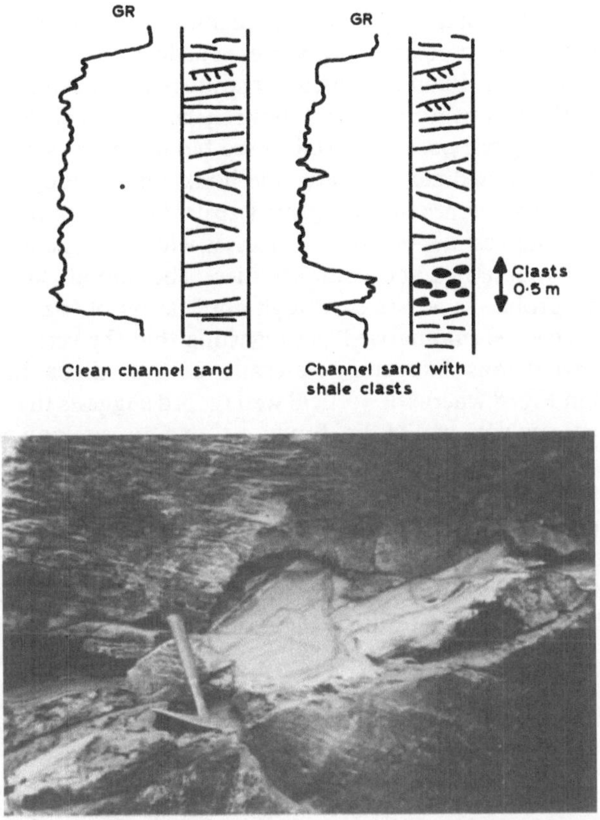

FIG. 7. Effect of shale clasts on sand body interpretation. Reproduced by permission of ERC Energy Resource Consultants Ltd.

Sedimentological investigation revealed that the 'shale' was a large overbank clast which was preserved in a channel sand. The clast is a highly localised event and should be ignored in the net sand estimation. The porosity, permeability and saturation estimates from core and log analysis apply to the net sand intervals. A geological model of the reservoir is required to interpret the relationship between observations of petrophysical properties at different wells in the reservoir.

3.3. Interwell Correlation of Petrophysical Characteristics

As stated at the beginning of this chapter, individual wells can sample only a small proportion of the reservoir. It therefore falls to a geological model to explain why the observed properties are found at particular wells, and to provide the basis for predicting properties at all locations between wells and in the reservoir. The main analytical techniques employed in this exercise are those of the sedimentologist and they rely heavily on observations from core. Wireline log response can also be used but it is necessary to have good core calibration. Sedimentological methods allow recognition of depositional energy regimes through grain size profiles and bedding structure. These in turn suggest depositional processes which, by modern analogues infer typical geometries and connectivities. The lateral extrapolation of vertical well profiles is possible through application of Walther's law[13] of facies which can be summarised as suggesting that the vertical record is a model of what was happening laterally at one single time. This is illustrated in Fig. 8 where the vertical well record suggests the depositional environments which would be found at one time along a coastline. Palaeogeographic reconstruction which respects all data observations has proved a valuable asset in guiding petrophysical property assignment in reservoir engineering simulation for field development studies (Fig. 9).

The basis for mapping porosity and thickness therefore lies in the recognition of a depositional process and perhaps also any diagenetic or structural events since deposition and before hydrocarbon emplacement,

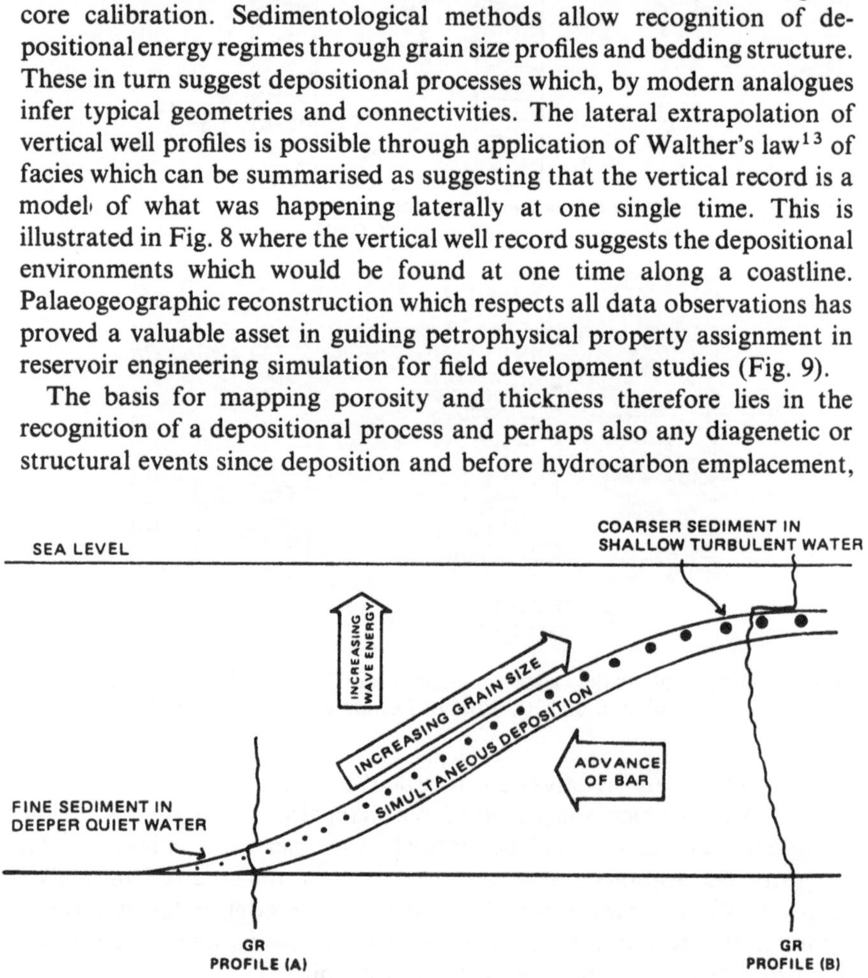

Note: The vertical sequence (at profile B) is the same as the lateral sequence being deposited at one time.

FIG. 8. Vertical and lateral sequence of deposition. (From Ref. 12. Reproduced by permission of ERC Energy Resource Consultants Ltd.)

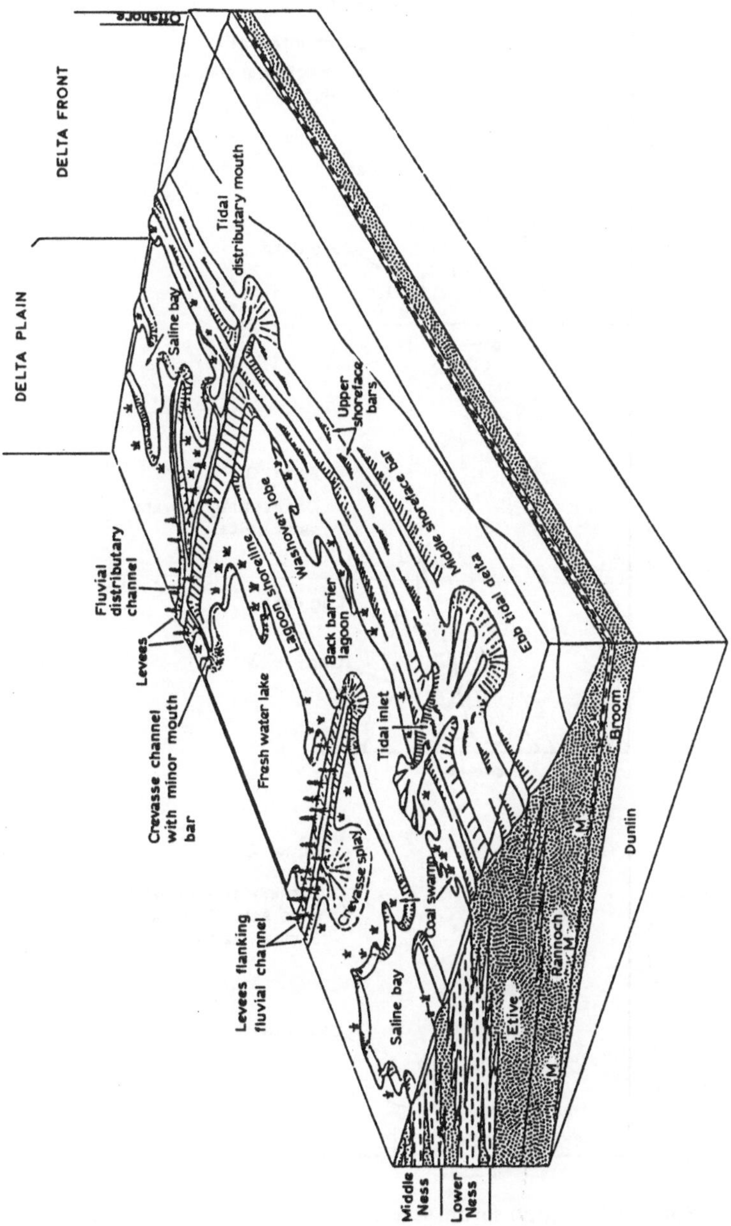

FIG. 9. Paleogeographic representations of the complex environments of the Ness, Etive, Rannoch and Broom interpreted from sedimentological analysis. (After Ref. 12. Reproduced by permission of ERC Energy Resource Consultants Ltd.)

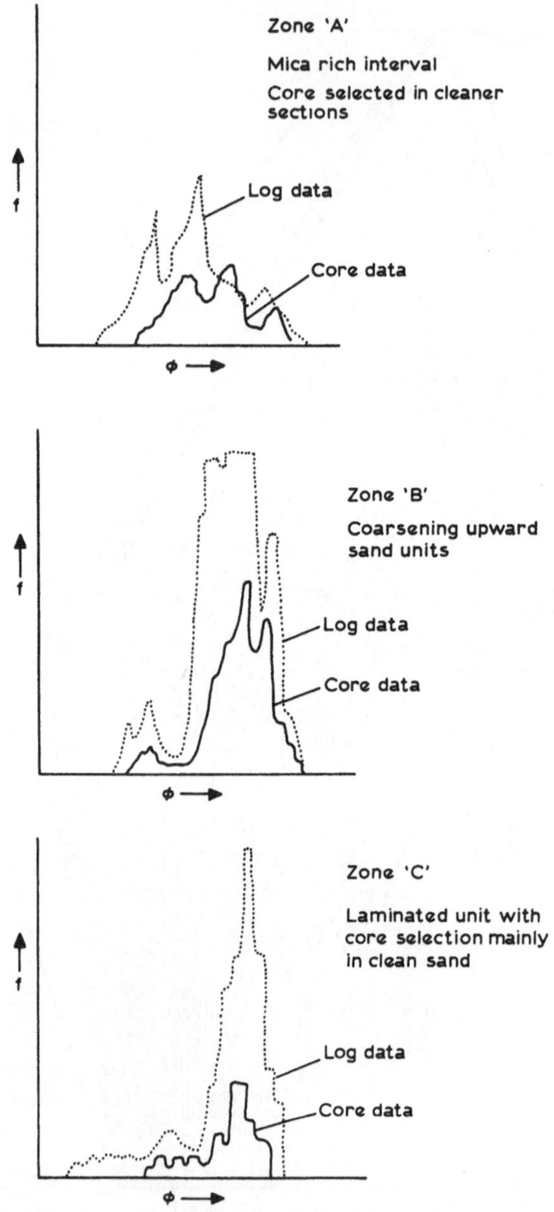

FIG. 10. Porosity histograms from several North Sea Field wells used in reservoir zonation and based on log interpretation and uncorrected routine core analysis.

and which have influenced pore space properties.[14–16] With thickness mapping it is necessary to interpret thickness trends in the light of geological models and with knowledge (or expectation) of the influence of faults. In many North Sea reservoir systems, fault activity during deposition as well as erosion has been responsible for crestal thinning.[17]

The recognition of individual zone character in a well and between wells is generally aided by analysis of histograms. The assumption inherent in the use of histograms is that a 'zone' contains a particular distribution of properties which make it 'heterogeneously homogeneous'.[18] The truncated normal distribution for porosity and the truncated log-normal distribution for permeability are frequently found in reservoir zones. Multimodel character in 'proposed' zones indicates that subzonation is required. Diagenetic modification to pore space can be depth controlled and/or saturation controlled in a region of a reservoir and is sometimes responsible for subzonation in a depositional unit.

Figure 10 shows log-derived and core-measured porosity histograms for a zone in a North Sea reservoir. That the core and log data are part of the same distribution can be checked using statistical tests. A non-parametric test of goodness of fit which has found application in petrophysical analysis is the Kolmogorov–Smirnov test.[19,20] For a series of data of length T and at a distance t_i from the start of the series to the ith event of a number of n events then the Kolmogorov–Smirnov statistic (KS) can be calculated as the absolute value of the larger of KS^+ or KS^- where

$$KS^+ = \max\left\{\frac{1}{n} - \frac{t_i}{T}\right\}\sqrt{n}$$

$$KS^- = \max\left\{\frac{t_i}{T} - \frac{(i-1)}{n}\right\}\sqrt{n}$$

A critical value of the statistic at the 5% level of significance is given by $(KS)_{crit} = 1{\cdot}36/\sqrt{n}$ and the null hypothesis is rejected at this level of significance if the test statistic (KS) exceeds this value. The test essentially compares the maximum deviation of an observed series (say the core porosity or log-derived porosity) from the 'theoretical' model for the zone, and in effect accounts for normalisation of the data set.

The zonation, both vertically and areally, of petrophysical properties must be guided by geological concepts. The interwell correlation of properties becomes particularly significant in describing reservoir continuity for hydrocarbon recovery calculations and geological ideas can, after some reservoir production, be augmented and tested with differential pressure drawdown data. The influence of geological concepts on reservoir mapping and correlation can best be illustrated by example.

First, consider a correlation between two wells, with well 1 located 1 km south-west and downdip of well 2. At the same distance above a well-established marker horizon well 1 and well 2 have oil sands with the characteristics given in Table 1.

Without further information it may be tempting to trend the thickness and petrophysical properties in a downdip deteriorating way as shown in Fig. 11(a). Let us now suppose that sedimentological study of core recovered from the wells suggests that the depositional character of well 1 is consistent with that of a crevasse splay or sheet flood event, whereas that of well 2 is more like that of a distributary channel. In terms of the relationship between the two sands a model such as that in Fig. 11(b) may be proposed, which shows a cross-section through the channel and its overbank flood facies.

It now becomes clearer that by recognising the depositional character and geometry of the sand body a more reasonable map could be prepared. Using an oriented core to help identify current direction in the channel we can draw Fig. 11(c) to represent contours of thickness and porosity (and permeability). Water saturation is a property dependent on pore geometry and permeability. The capillary pressure curves for air–brine measurements using cores taken at position C (Fig. 11(b)) in each well are as shown on Fig. 11(d) and show that rock-type zonation may be necessary. The capillary pressure data can be normalised using a Leverett *J*-function approach to check this zonation. In the example shown in Fig. 11(e) the plot of S_w against $J_{(sw)}$ where

$$J_{(sw)} = \frac{P_{c(sw)}}{\sigma \cos \theta} \sqrt{\left[\frac{k}{\phi}\right]}$$

suggests that the distinction in zonation may not be significant.

A second example can be used to illustrate some alternatives in mapping which lead to uncertainty in hydrocarbon in-place estimation. The

TABLE 1

	Sand thickness (m)	*Sand grain size, character*	*Mean ϕ*	*Mean k (mD)*	*S_{w_i} (fraction)*
Well 1	2	f-m, poor sorting	0·20	50	0·38
Well 2	10	f-c, fining upwards	0·25	2 000	0·20

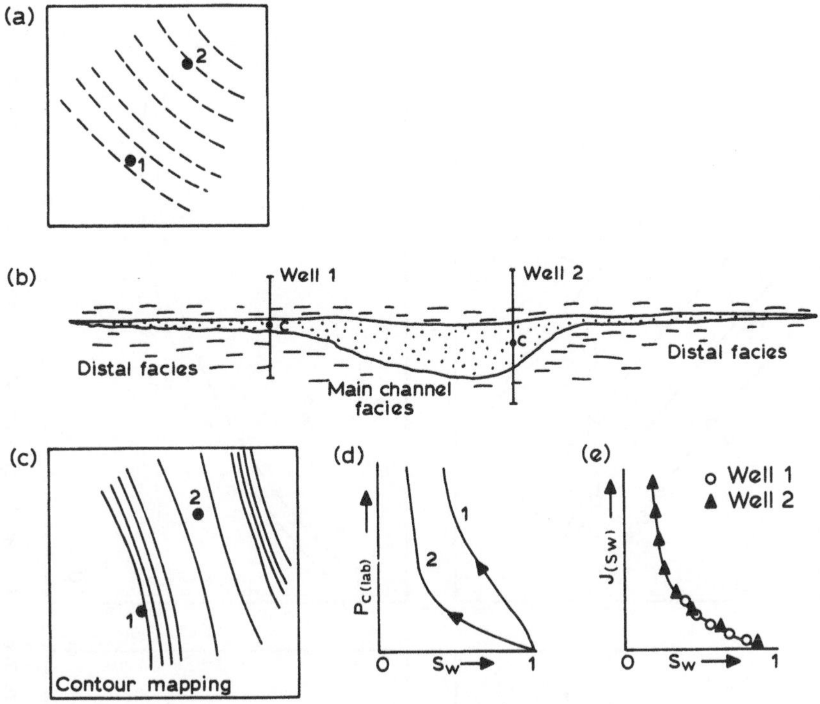

FIG. 11. Well correlation: (a) contour mapping 1; (b) cross-section; (c) contour mapping 2; (d) capillary pressures; (e) *J*-function correlation.

uncertainty is, however, reduced as either more information is acquired or more detailed geological modelling work is performed with the available core material. As before, the example considers two wells with the porosities, permeabilities, saturations and thicknesses indicated in Table 2.

Consideration of the reservoir delineation at this time shows that there is a combination of fault and fluid closures with well A being in an aquifer as

TABLE 2

	Sand thickness (m)	*Sand mean porosity (fraction)*	*Sand mean permeability (mD)*	S_{w_i} *(fraction)*
Well A	50	0·20	300	1·0
Well B	48	0·25	3 000	0·2

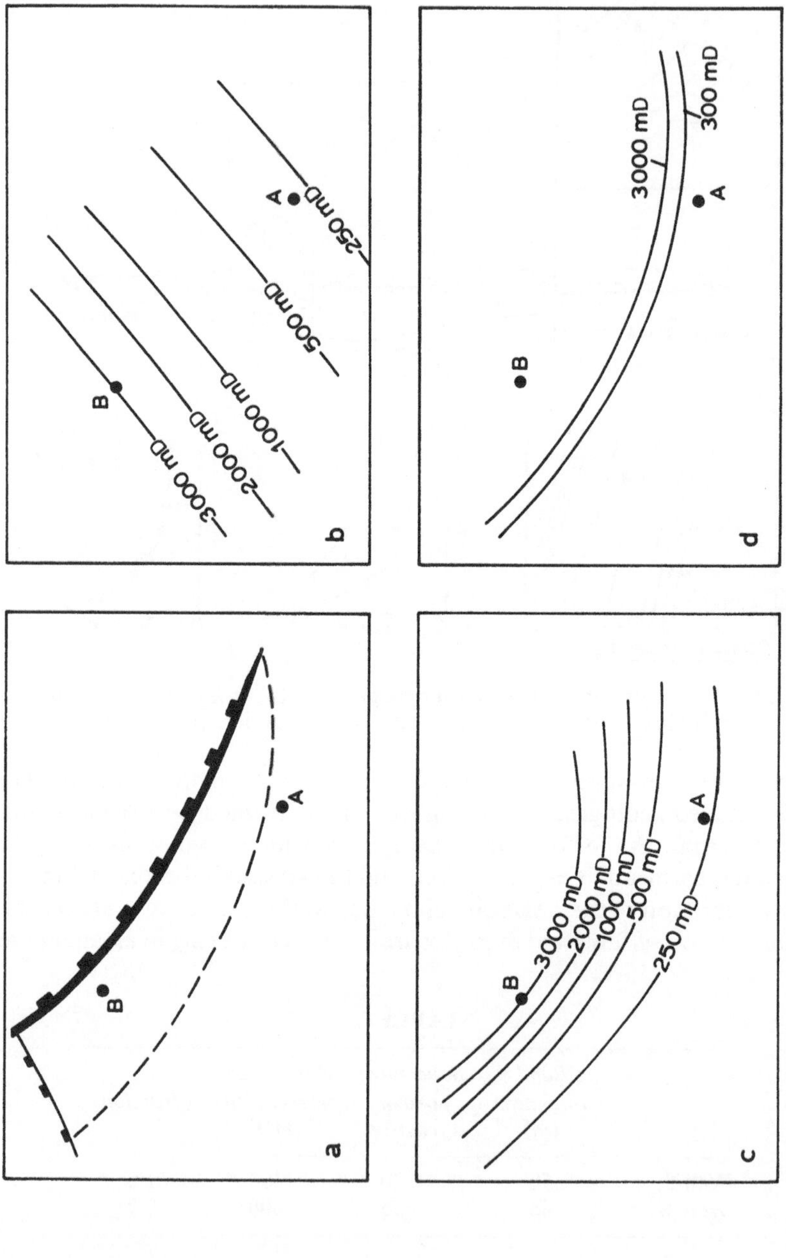

FIG. 12. Petrophysical mapping: (a) well location; (b) permeability map 1; (c) permeability map 2; (d) permeability map 3.

shown in Fig. 12(a). A number of alternative models can be proposed to map porosity and permeability. The permeability is assumed to be related to porosity and will be used in this example. First, imagine that an indication of depositional dip has suggested that permeability should be mapped according to the depositional model; then Fig. 12(b) might result. Note that a logarithmic-based contour interval has been chosen for permeability. If evidence suggested that the present structure had been an influence during deposition and that the sands were of better quality updip, then Fig. 12(c) might result. With evidence from mineralogical and diagenetic analyses we might introduce a further set of models, one of which is illustrated in Fig. 12(d). In this model the concept that authigenic clays have developed in the pore space filled with water and not in that presently occupied by hydrocarbon is proposed. The sand above the transition zone is then unaffected by the diagenetic changes.

The influence of geological concepts as well as data validity has been shown to have significance in the representation of hydrocarbon in place. Through recognition of uncertainty and by subjective quantification of the probabilistic values of independent parameters which combine to define hydrocarbon volume, we obtain a more realistic basis for reservoir evaluation than can be provided from deterministic estimates.

4. RECOVERY FACTOR AND RECOVERY MECHANISMS

The definition of recovery factor used in this chapter is the ratio between cumulative hydrocarbon recovered at surface conditions of temperature and pressure and the initial volume of hydrocarbon in place at the same surface conditions. In terms of oil measured at stock tank conditions, the recovery factor, RF, at some time t is

$$\mathrm{RF} = \left.\frac{N_\mathrm{p}}{N}\right]_t$$

where N_p is the cumulative oil produced and N is the initial volume of oil in place.

The association of recovery factor with time is important since it introduces the idea that reservoir management can play a part, and that a decision to abandon production will perhaps be primarily an economic decision for the facilities available. The definition of an ultimate recovery factor rather than an economic one requires a basis of unit pore volume.

For an oil reservoir of initial average water saturation $\bar{S}_{w_i}$ and average residual oil saturation $\bar{S}_{or}$:

$$RF_{(ult)} = \frac{1 - \bar{S}_{w_i} - \bar{S}_{or}}{1 - \bar{S}_{w_i}}$$

It may be apparent that the reduction of oil saturation to an overall average residual value $\bar{S}_{or}$ requires production of 100 per cent water at abandonment. The definition of the value $\bar{S}_{or}$ is in itself problematical since it may assume contact of displacing phase fluid with oil in all parts of the reservoir (i.e. 100% sweep), and an ability to define possibly different residual oil saturations in each reservoir zone. The calculation of average initial water saturation $\bar{S}_{w_i}$ which is also a zone-averaged value depends also on zonation ability.

4.1. Probabilistic Representation of Recoverable Reserves

Since the recovery factor (RF) is represented as the fraction of initial hydrocarbon in place that has been recovered, the 'real' value of the initial hydrocarbon in place is of less significance mechanistically in the calculation of RF. The recovery factor may thus represent a normalised calculation for a range of probabilistic values of initial hydrocarbon in place. The volume of hydrocarbon finally produced or the recoverable reserve (RR) will, however, result from an association of recovery factors with the probabilistic hydrocarbon in-place calculation. Using the terminology from Section 2, an appropriate representation in the jth zone of a reservoir is:

$$RR_j(p1, p2, p3, p4, p5) = \frac{A_j(p1)h_{N_j}(p2)[\phi S_{hc}]_j(p3)[RF]_j(p5)}{[B_{oi}]_j(p4)}$$

where h_{N_j} is the net thickness of the jth zone and $[B_{oi}]_j$ is the initial oil formation volume factor in the jth zone. The range of recovery factors calculated are then generally normalised in a cumulative distribution as shown in Fig. 1.

The probabilistic representation of the recovery factor arises from consideration of (i) recovery mechanisms and (ii) number and location of wells and completions, aquifer description and process constraints.

It is not within the scope of this chapter to review all the mechanisms which allow recovery of hydrocarbons. For the principles many specialised texts[21-26] are available: suffice to say here that processes can be considered as leading to recovery by expansion or displacement and that displacement can be predominantly miscible or immiscible. In addition, the recovery processes involve capillary, viscous and gravity force balances and can be analysed on a compositional or phase (oil, water, gas) basis. The hydrocarbon remaining in a reservoir after abandonment of a recovery process will be distributed in the reservoir system according to pore geometry, dynamic pressure gradients and sweep efficiency of displacement. The evaluation of recovery efficiency therefore requires consideration of reservoir heterogeneity and of rock and fluid dependent properties. The representation of reservoir heterogeneity is essentially a case of matching zonation to the scale of the reservoir calculation necessary. For example, calculation of reservoir recovery at a number of equilibrium pressure stages during exploitation could be achieved with a zero-dimensional model (also called a General Material Balance). In such a calculation the difficulty lies in representing an average hydrocarbon zone pressure and in defining pore volume and average saturation. In addition, the link between average well performance and average saturation requires, at present, a functional empirical relationship. Any trend away from simplified reservoir description and analytical solutions (even with pseudo-functions) ends up with the application of numerical reservoir simulation models. There are several texts on the formulation and development of reservoir simulation codes.[27–30] Most of these highlight the method of formulating the pressure and saturation dependent terms linking the non-linear differential equations of flow and continuity and they outline approximate solution methods. The texts consider radial and cartesian co-ordinate systems for defining the reservoir flow geometry and, in general, outline how empirical terms such as relative permeabilities can be treated. The validity of a reservoir simulation model appears to be that with particular reservoir petrophysical description and well characterisation the essential pressure-rate-saturation historical performance of producing reservoirs can be reproduced by the model. The degree of discretisation of a reservoir in a simulation exercise is within the choice of a reservoir engineer and is in practice limited only by computational costs—it therefore resolves to an economic assessment of the value of the result and the time to acquire it.

For a given recovery mechanism, for example oil displacement by water injection, the number and positioning of wells and their completion intervals will possibly affect production rates and volumes. These variables with their inherent well constraints on say, water cut, minimum bottom hole pressure, etc., will lead to a range of recovery factors. Further uncertainty however will relate to reservoir continuity and zonation and the influence of rock and fluid dependent properties.

4.2. Rock and Fluid Dependent Properties

Geological and petrophysical models are the basis for representation of reservoir continuity and zonation and the subsequent calculation of recoverable hydrocarbons. It is almost impossible for anyone having looked closely at outcrop sections of reservoir quality sedimentary rocks to imagine such a thing as a homogeneous reservoir. In reservoir simulation models, homogeneous property cells, maybe as small as ten thousand cubic metres in rock volume, are defined. There is no concept of pressure or saturation distribution with a given cell, and flow between cells across the intercell boundary is usually calculated. The representation of transmissibility between reservoir simulator cells at the boundaries between them is an important task for definition by a geological model. Transmissibility in a given direction is usually represented by a grouping of equivalent terms from the linear Darcy equation. In the X-direction plane, the transmissibility T_X is given with reference to oil flow in Fig. 13 as follows:

$$T_X = \frac{q_O \mu_O}{\Phi_1 - \Phi_2} = \frac{2A(k_H k_{ro})_B}{(L_1 + L_2)}$$

where Φ_1 and Φ_2 are the effective potentials (datum corrected pressures) in cell 1 and cell 2 respectively, q_O is a volumetric reservoir condition rate of flow, k_H and k_{ro} are the harmonic average permeability and the relative permeability at the intercell boundary, respectively, A is the cross-sectional area at the boundary, L_1 and L_2 are the cell lengths in the X-direction of cell 1 and cell 2, respectively, μ_O is the oil viscosity, and B is a subscript indicating a boundary value at the interface between cells.

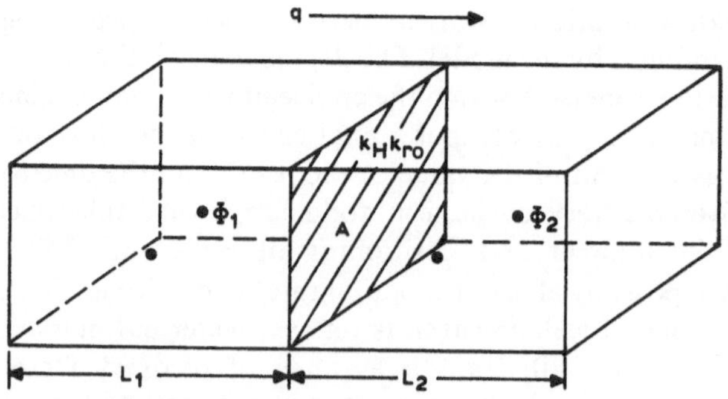

FIG. 13. Transmissibility in cell models.

To represent a flow barrier or flow restriction at an intercell boundary, i.e. a fault or a shale, the transmissibility can be multiplied by a factor which can be between zero and unity. For representation of a simplistic fracture the factor could be greater than unity. The factors are generally obtained through a history match of observed reservoir performance with calculated performance. Matching of vertical pressure distributions obtained from RFT tests on new wells in producing reservoirs provide valuable guidance on the magnitude of vertical transmissibility factors.[18]

The permeability at a cell boundary is normally calculated as the harmonic average between cell values. The relative permeability is most frequently an upstream value from cell (or cells) at the start of a timestep of calculation. More recently dynamic weighting of relative permeability values has been applied.[31]

The sources of thickness, porosity, saturation and permeability data for use in reservoir zonation have been discussed in an earlier part of this chapter. The reservoir zonation and interzone transmissibility are based on geological concepts and pressure distributions. It is intuitively reasonable to expect that reservoir zones which are recognised by pore size and pore size distribution differences will have different relative permeability and residual saturation character. In addition the zones should exhibit different dynamic capillary pressure character and all these functions will be dependent on the particular displacement process. At present, these data require laboratory determination.

Relative permeability measurements for oil–water systems at reservoir conditions of temperature and pressure and using reservoir fluids have been demonstrated to give different results from experiments with the same rock using room conditions and refined fluids.[32,33] Some displacement processes which are believed to occur in a reservoir are very difficult to reproduce under laboratory conditions. For example, updip reinjection of separator gas at 'miscibility' pressure into an oil zone will lead to high recovery efficiency in laboratory tests.[34] In the field, the process may be more complex with multicontact miscibility occurring at the leading edge but with poor sweep efficiency being enhanced by reservoir heterogeneity. Under such circumstances gravity-controlled gas movement could lead to override of oil and consequent earlier breakthrough of gas than expected at downdip producing wells. The inability to model the displacement process in reservoir simulators arises from the uncertainty in the miscibility mechanisms and in the description of residual saturations, relative permeabilities and, particularly, in transmissibilities.

Relative permeability data are generally shrouded in uncertainty with respect to their derivation, application and representation. In reservoir engineering calculations they are used in homogeneous regions of the reservoir and may through translation to static, dynamic or well function pseudo-relative permeabilities be used to provide the performance of a more complex system. The source of relative permeability data is from laboratory tests in core plugs, simulating the displacement process. That a core plug represents a given reservoir zone in petrophysical properties is usually assumed. It may perhaps be further assumed (possibly erroneously!) that the wetting character of the rock with respect to water and hydrocarbons has not been changed by the coring, recovery, storage, cleaning and preparation processes and that the fluids used in the laboratory tests are equivalent to those in the reservoir at reservoir conditions. The core plug test models diffuse flow conditions with virtually 100 per cent sweep and can be an unsteady state or steady state displacement process. The relative permeability concept is still empirical[35] in its application and traditionally relates two-phase system data to the assumed wetting phase saturation (certainly a particularly significant parameter) and the direction of saturation gradients.[36,37] More attention is now being paid to understanding more complex secondary relationships, particularly with pore geometry and pore filling materials.

Three-phase relative permeability data are experimentally sparse and even more difficult to relate to rock properties than two-phase data. Representation of three-phase relationships by linking sets of two-phase data have been proposed in the literature and have found application in calculations.[38–42]

Residual saturations depend on viscous, capillary and gravity force ratios, in particular pore geometry systems with defined wetting preference. The laboratory data tend to use capillary number ($V\mu_D/\sigma \cos \theta$) relationships to correlate residual saturations in core plugs but scaling criteria, as yet unproven, must be used to translate the result to reservoir engineering field calculations. In the equation, V is the displacing phase velocity (volumetric flow rate/area), μ_D is the displacing phase viscosity, σ is the interfacial tension and θ is the contact angle.

5. CONCLUSION

The sensitivity of reservoir management plans to reservoir description uncertainty is a fundamental aspect of reservoir engineering. The representation of uncertainty through definition of probabilistic distributions of recovery factors, and through combination with volumetric hydrocarbon in-place estimates leads to a time-dependent probabilistic representation of recoverable reserves.

REFERENCES

1. ERC Energy Resource Consultants Ltd (1982) Petroleum Consultants Report to the Secretary of State for Energy re: the Petroleum Interests of Britoil plc, Dated November 10th, 1982, and published in the UK national press (*Times*, *Financial Times*, etc.) Friday November 12th, 1982.
2. Martinez, A. R. and Ion, D. C. (1984) Classification and nomenclature systems for petroleum and petroleum reserves. Study Group Report, in *Proc. Eleventh World Petroleum Congress*, Vol. 2, J. Wiley and Sons, London and New York, p. 325.
3. Walstrom, J. E., Mueller, T. D. and McFarlane, R. C. (1967) Evaluating uncertainty in engineering calculations. *J. Pet. Tech.*, **19**, 1595.
4. Newendorp, P. D. and Root, P. J. (1967) Risk analysis in drilling investment decisions. Paper No. 1932, *42nd Ann. Fall Meeting Soc. Petrol. Eng.*, Houston, Texas. Reprinted in *Soc. Pet. Eng. Pet. Trans.*, Reprint Series, No. 3 (1970), p. 188.
5. Heaviside, J., Langley, G. O. and Pallatt, N. (1983) Permeability characteristics of Magnus rcservoir rocks. In *Proc. Eighth European Formation Evaluation Symposium*, Soc. Prof. Well Log. An., London.
6. Monicard, R. P. (1980) *Properties of Reservoir Rocks: Core Analysis*, Editions Technip, Paris.
7. Lombos, L. (1983) Porosity measurement of core samples. Technical Information Note, Robertson Research plc.
8. Hook, J. (1983) The precision of core analysis data and some implications for reservoir evaluation. In *Proc. 24th Ann. Logging Symp.*, Calgary, Canada, Soc. Prof. Well Log. An.
9. Bilsland, M. and Archer, J. S. (1984) Research Note, Mineral Resources Engineering Department, Imperial College of Science and Technology, London.
10. Juhasz, I. (1981) Normalised Q_v—the key to shaley sand evaluation using the Waxman–Smits equation in the absence of core data. In *Proc. 22nd Ann. Logging Symp.*, Soc. Prof. Well Log. An., New Orleans, Louisiana.
11. Archer, J. S. and Hancock, N. J. (1980) An appreciation of Middle Brent sand reservoir features by analogy with Yorkshire coast outcrops. In *Proc. European Offshore Petroleum Conference*, London, Soc. Pet. Eng., Paper 197, Vol. 1, p. 501.
12. Archer, J. S. and Hancock, N. J. (1979) An integrated field course in geology and reservoir engineering: North Sea deltaic reservoirs. RRI/ERC manual.
13. Walther, J. (1894) *Einleitung in die Geologie als Historische Wissenschaft Bd. 3, Lithogenesis der Gegenwart*, Fischer-Verlag, Jena.

14. WEBER, K. J. (1982) Influence of common sedimentary structures on fluid flow in reservoir models. *J. Pet. Tech.*, **34** (March), 665.
15. DELANEY, R. P. and TSANG, P. B. (1982) Computer reservoir continuity study at Judy Creek. *J. Can. Pet. Tech.*, **21** (Jan–Feb), 38.
16. HANCOCK, N. J. (1978) Diagenetic modelling in the Middle Jurasic Brent Sand of the Northern North Sea. In *Proc. European Offshore Petroleum Conference*, London, Paper 92, Vol. 2, p. 275.
17. VAN RIJSWIJK, J. J., ROWBOTTOM, D. J. and SPRAKES, C. W. (1980) The Dunlin Field: A review of field development and reservoir performance to date. In *Proc. European Offshore Petroleum Conference*, London, Paper 168, Vol. 1, p. 217.
18. ARCHER, J. S. (1984) Reservoir definition and characterisation for analysis and simulation. In *Proc. Eleventh World Petroleum Congress*, Vol. 3, PD(6), J. Wiley and Sons, London and New York, p. 65.
19. MILLER, R. L. and KAHN, J. S. (1962) *Statistical Analysis in the Geological Sciences*, J. Wiley and Sons, Inc., New York.
20. DAVIS, J. C. (1973) *Statistics and Data Analysis in Geology*, J. Wiley and Sons, London and New York.
21. DAKE, L. P. (1978) *Fundamentals of Reservoir Engineering*, Elsevier Scientific Publishing Co., Amsterdam.
22. CRAFT, B. C. and HAWKINS, M. F. (1959) *Applied Petroleum Reservoir Engineering*, Prentice-Hall, Englewood Cliffs, New Jersey.
23. CRAIG, F. F. (1971) *The Reservoir Engineering Aspects of Water Flooding*, Monograph 3, Soc. Pet. Eng., Dallas.
24. SMITH, C. R. (1975) *Mechanics of Secondary Oil Recovery*, Reinhold, New York. Reprinted by Kreiger Publishing Co., New York.
25. SANDREA, R. and NIELSEN, R. F. (1974) *Dynamics of Petroleum Reservoirs under Gas Injection*, Gulf Publishing Co., Houston, Texas.
26. ARCHER, J. S. and WALL, C. G. (1985) *Principles of Petroleum Engineering*, Graham and Trotman, London (in press).
27. AZIZ, K. and SETTARI, A. (1979) *Petroleum Reservoir Simulation*, Applied Science Publishers, London.
28. PEACEMAN, D. W. (1977) *Fundamentals of Numerical Reservoir Simulation*, Elsevier Scientific Publishing Co., Amsterdam.
29. CRITCHLOW, H. (1977) *Modern Reservoir Engineering: A Simulation Approach*, Prentice-Hall, Englewood Cliffs, New Jersey.
30. THOMAS, G. W. (1981) *Principles of Hydrocarbon Reservoir Simulation*, International Human Resources Development Corporation (IHRDC), Boston, Mass.
31. WILSON, D. C., TAN, T. C. and CASINADER, P. C. (1981) Control of numerical dispersion in computer simulation. In *Proc. 1981 European Symposium on Enhanced Oil Recovery*, Elsevier Scientific Publishing Co., Amsterdam, p. 425.
32. COLPITTS, G. and HUNTER, D. (1964) Laboratory displacement of oil by water under simulated reservoir conditions. *J. Can. Pet. Tech.*, **3**(2), 66.
33. CUIEC, L., LONGERON, D. and PACSIRSZKY, J. (1979) On the necessity of respecting reservoir conditions in laboratory displacement studies. Paper SPE 7785, *Middle East Oil Tech. Conf.*, Soc. Pet. Eng.
34. BATH, P. G. H., VAN DER BURGH, J. and YPMA, J. G. (1984) Enhanced oil recovery in the North Sea. In *Proc. Eleventh World Petroleum Congress*, Vol. 3 (RTD2), J. Wiley and Sons, London and New York, p. 189.
35. BROWN, C. E. and LANGLEY, G. O. (1981) The provision of laboratory data for

EOR simulation. In *Proc 1981 European Symposium on Enhanced Oil Recovery*, Elsevier Scientific Publishing Co., Amsterdam, p. 81.
36. KILLOUGH, J. E. (1976) Reservoir simulation with history dependent saturation functions. *Soc. Pet. Eng. J.*, **16**, 37.
37. CARLSON, F. M. (1981) Simulation of relative permeability hysteresis to the non-wetting phase. Paper SPE 10157, Soc. Pet. Eng., *Proc. 56th Ann. Fall Meeting*, San Antonio, Texas.
38. COREY, A. T. *et al.* (1956) Three phase relative permeability. *Pet. Trans. AIME*, **207**, 349.
39. STONE, H. L. (1973) Estimation of three phase relative permeability and residual oil data. *J. Can. Pet. Tech.*, **12**(4), 53.
40. LAND, C. S. (1968) Calculation of imbibition relative permeability for two and three phase flow from rock properties. *Soc. Pet. Eng. J.*, **8**, 149.
41. DIETRICH, J. K. and BONDOR, P. L. (1976) Three phase oil relative permeability models. Paper SPE 6044, Soc. Pet. Eng., *Proc. 51st Ann. Fall Meeting*, New Orleans, Louisiana.
42. FAYERS, F. J. and MATTHEWS, J. D. (1984) Evaluation of normalised stone's method for estimating three phase relative permeabilities. *Soc. Pet. Eng. J.*, **24**, 224.
43. PALLAT, N., WILSON, J. and MCHARDY, W. (1984). The relationship between permeability and morphology of diagenetic illite in reservoir rocks. *J. Pet. Tech.*, **36** (December) 2225.
44. GARB, F. (1985) Oil and gas reserve classification, estimation and evaluation. *J. Pet. Tech.*, **37** (March) 373.

Chapter 3

ADVANCES IN FORMATION EVALUATION

P. THREADGOLD*
Tesel plc, Basingstoke, UK

1. INTRODUCTION

The origin of formation evaluation was almost certainly based on examination of cores and cuttings, and on measurement of producing pressures. Resistivity measurements downhole were introduced by the Schlumberger brothers in the mid 1920s. It was with some justification that the method was labelled 'electrical coring'. Electrical or wireline logging gained acceptance in the petroleum industry in the late 1940s, from which time it has developed into a major discipline, and now effectively forms the basis of formation evaluation.

In the early days, log measurements were calibrated by correlation with core measurements, particularly in the case of the derivation of porosity. It was some considerable time before the measurements attained sufficient accuracy and reliability to be used in their own right. Discrepancies between log and core data were always considered to result from errors in the log-derived data. While this was very largely true, it is only relatively recently that it has been generally accepted that core data even when accurate, are often statistically unrepresentative of the reservoir rock. In addition, stress relief, etc., during drilling and coring can cause significant changes in the physical nature of the core as it is cut and depressured when brought to the surface. Modern technology, in particular data processing, has brought about a more accurate definition of logging response characteristics, resulting in more reliable corrections for environmental

* Present address: Coombe Cottage, Overton, Nr. Port Eynon, Gower, West Glamorgan SA3 1NR, UK.

conditions. The large multiplicity of logging measurements now available allows cross-checking of data and more rigorous solutions to the various equations relating tool response to formation characteristics. Modern wireline logging is the only method of obtaining continuous information over the whole depth of a borehole, which directly relates to the *in-situ* characteristics of the rock penetrated.

The running of modern wireline logs is fast and efficient. The derived data are universally understood and can be used by petroleum engineers, geologists, geophysicists, drillers, etc., in their daily decision-making. Reservoir descriptions based on log data are also universally used in making management decisions on exploitation of the reservoir.

Log data are now sufficiently accurate and reliable to enable sound practical decisions to be made in appraisal of exploratory boreholes, and for development of reservoirs in the majority of cases. Much, however, remains to be resolved. The reservoir rock is a statistically complex mixture of components, and reservoir descriptions, while essentially valid, are often grossly oversimplified. This was not too important for the routine type of reservoir exploited over past-history. However, as is often stated: 'the easy oil has nearly all been found'. The next decade will see the serious exploitation of difficult sources of hydrocarbons, tar sands, heavy oil deposits, shale oil, etc., and a larger quantity of accurate *in-situ* data will be required.

In addition, the coal, minerals and civil engineering industries all use *in-situ* measurements to a greater or lesser extent. These industries have specific problems which will require their own solution. It is certain that more effort will be put into these diverse problems in the future, and that the results will be integrated into formation evaluation as a whole. The tremendous advances of the last decade or so, will be matched or even surpassed in the next.

2. INTERPRETATION TECHNIQUES

The work of Archie, which related to homogeneous porous media having an electrically inert solid matrix, still remains the basis of modern open hole log interpretation methods. Considerable advances, however, have been made in the last decade. These are the result, first, of the increasing application of computer technology to well log data acquisition and handling and, secondly, of a substantial effort by a variety of workers on the relevant physical characteristics of clay materials and their effects on well log response.

2.1. Computer Technology in Well Logging

The application of computer technology to well log data processing has made it possible to deal with large volumes of data efficiently and quickly. This has resulted in detailed complicated interpretation equations being solved for large suites of logging runs, at regular small depth increments (often $\frac{1}{4}$ m) over zones of interest in the borehole.

In addition, the introduction of process control type techniques to the data-acquisition phase of well logging has resulted in an increase in accuracy, speed and reliability. The amount of data available from the logging surveys is increasing almost daily. This increase in speed and efficiency, particularly in critical situations, is of considerable financial value to the operating oil company.

Initially, computers were used at base for detailed log interpretation, thus requiring digitisation of the conventional analogue records. The development and introduction of more complex logging tools by the service companies, in particular the dipmeter and the different combination tools, posed various data recording problems. These were solved by the introduction of digital recording on magnetic tape. Introduction of digital recording of all logs followed, initially using analogue-to-digital recording. The conversion to full digital recording and data handling at the well site by the logging companies completed its initial stage at the end of the 1970s and development is proceeding very quickly. It is now possible to carry out initial log interpretation at the well site using fairly complex data handling and interpretation methods. This produces a 'first-look' result which is, in fact, usually more accurate and more detailed than the early long-term log analysis methods carried out manually.

In addition to the on-site interpretation, the digital log data can be transmitted via radio and telephone links to an interpretation base, during and immediately following the log data acquisition at the well site. Use of satellites for log data transmission over continents is also an available service. At the well site, digital-to-analogue converters still enable conventional analogue presentations to be produced in parallel with the digital tape record. This enables the logging engineer, the geologist, and/or the petroleum engineer at site, to monitor the logging process in the conventional manner. However, the on-site computer plays a further role in that it is programmed to calibrate and check each of the logging tools used, thus relieving the logging engineer of a large amount of relatively boring routine work, which formerly was prone to human error, particularly during long tedious logging surveys in excess of 24 h. The role of the client log specialist at site is also relieved of much of the drudgery of the on-site work involved in witnessing the logs and carrying out some form of quick look manual interpretation. The log specialist now can work with the logging engineer using the on-site computer unit to produce log data cross plots for examination, and to assist in the choice of interpretation parameters to use in an on-site interpretation run.

Most major operators in the petroleum industry have, or have access to, computing facilities. The facilities available with the modern, relatively cheap, minicomputers are larger than those of an expensive large digital computer of a decade earlier. This means that the function of the log specialist at base has changed quite positively. The work is ceasing to be on single well log evaluation, and increasingly concentrates on detailed multiple well evaluation for whole reservoir units. The work of the reservoir geologist, the petrophysicist (log specialist) and the reservoir/petroleum engineer has therefore become much more closely linked. This is resulting in the development of proper integrated formation evaluation procedures. The immediate benefit is a much better and more comprehensive understanding of the nature of the reservoir, resulting in more positive practical definitions of the real development and production problems. Advances in technology can rarely be isolated from the overall system. There is always a 'knock-on' iterative effect. In the case of formation evaluation and downhole data acquisition, the better definition of problems, generates improvements in, and development of, the basic data acquisition systems by the logging service companies. It could be argued, therefore, that the major advance in the last decade, is the integration of the various aspects of formation evaluation into a more unified whole.

2.2. The Clay Problem

The clay problem, simply stated, is that the inclusion of clay or shaly materials as a constituent of reservoir rock affects the response of all conventional logging tools to a greater or less extent. The effect for a specific tool depends on the volume fraction of clay, the type of clay, and in many cases the nature of the environment. The response of the tool to the clay needs to be computed, in order to allow for it in the interpretation. The problem is to define the effect of the clay on the tool in question, and at each interpretation depth or interval, to compute the volume proportion of clay present. Work on this has been continuous since the early 1950s, since which time steady improvements have been made both in the understanding of the physical nature of the problem, and in methods of allowing for the clay effect on the log data responses. The ability to handle large amounts of data easily and quickly using computers noticeably speeded these improvements. The increasing similarity between the more up to date of the various hypotheses concerning the nature of the clay effect on various logging parameters, particularly on formation conductivity, suggests that a consistent physically sound solution to the problem is very close.

A reservoir rock is one having porosity between the basic rock-forming minerals, that is available for the storage of the interstitial reservoir fluids. In addition, all or a substantial portion of the porosity must be interconnected to allow movement of fluids into and out of the rock. When clays are present, part of this total porosity is occupied by the clay, leaving a reduced fraction of the total porosity available for storage of fluids. This fraction is the effective porosity. The distribution of the clay within the reservoir can also vary. It can exist as discrete particles of clay dispersed within the porosity. It can coat the original grain surfaces, or it can exist as continuous clay or shale laminae forming very low (almost zero) permeability barriers to flow. Clearly, clay reduces the flow characteristics of the rock, depending on the volume and distribution present.

In the context, therefore, of estimating the reserves present in a reservoir, the first required parameter is the pore space available for fluid storage, which is the effective porosity ϕ_e. This is the difference between the total (intergranular) porosity ϕ_T, and the volume of clay present V_{cl}, i.e.

$$\phi_e = \phi_T - V_{cl}$$

The response of each of the three basic porosity-dependent logging devices, the density, acoustic velocity and neutron logging tools, are the resultant sum of their responses to the component parts of the reservoir. Hence, for the density tool, the bulk density ρ_b is the sum of the densities of the component parts, i.e.

$$\rho_b = \sum_{i=1}^{i=n} V_{ma_i}\rho_{ma_i} + V_{cl}\rho_{cl} + (\phi_e\rho_f)$$

where

$$\sum_{i=1}^{i=n} V_{ma_i} + V_{cl} + \phi_e = 1$$

and where V_{ma_i} is the volume fraction of rock matrix type i, ρ_{ma_i} is the density of the rock matrix i, V_{cl} is the volume fraction of clay, ρ_{cl} is the clay density, ρ_f is the density of the pore-filling fluid and ϕ_e is the effective (fluid-filled) porosity. This means that the number and volume proportions of the components of the rock matrix need to be defined, e.g. calcite, dolomite, quartz, and their densities known. V_{cl} and ρ_{cl} need defining, as also does ρ_f. It must be remembered that the fluid fraction may be any proportion of gas, oil or water.

Similar equations may be written for the acoustic velocity and neutron log responses. Using the three equations and the unity equation, only four unknowns can be resolved uniquely. Independent evidence is therefore obtained from cuttings descriptions and previous local knowledge, etc., to define the nature of the rock matrix. Other logs such as natural γ, spontaneous potential, resistivity, thermal neutron die away, also respond to clay. A series of clay indicators proposed by Poupon and Gaymard (1970) and the type of analysis developed, forms the basis of estimation of the clay volume present.

For initial analysis, as the porosity-sensitive devices have only a shallow-to-moderate depth of investigation, the pore-filling fluid is assumed to be invaded zone mud filtrate. However, in hydrocarbon-bearing formations, residual hydrocarbon will always be present. The initial porosity estimate is used in conjunction with the shallow investigation resistivity devices to obtain an estimate of the filtrate saturation S_{xo} and the residual hydrocarbon saturation S_{hc} in the invaded zone. The pore-filling fluid characteristics are adjusted accordingly and an iteration process carried out. Clearly a knowledge of the nature of the hydrocarbons (oil or gas) is also required, as is also the effect of oil and gas on the log responses. The basic work on this aspect of the problem was carried out by Poupon and Gaymard (1968). Considerable investigation has also been carried out into the response characteristics of the various logging devices to the many minerals which may form the rock matrix. The publications of Edmunson and Raymer (1979) and Wells *et al.* (1979) give valuable data in this respect.

While an accurate assessment of effective porosity in formations with complex lithology is difficult, it can in most cases be satisfactorily achieved. However, the effect of clay on the resistivity devices and the subsequent computation of the water saturation fraction has proved a much more difficult problem. It would appear that at the present time the correct solution to the problem is very close, if not already available. One of the difficulties has been historical, in that clay was rather loosely termed 'conductive solid', which was quite untrue as many early workers had demonstrated. (Dry clay has an electrical conductivity approaching zero.) The electrical characteristics of wet clay depend on the environmental conditions, and in particular on the ionic concentration of the fluid in equilibrium with the clay. Essentially, clay particles have an unsatisfied surface charge (negative), which is counterbalanced by the counter cations (positive) of the equilibrium fluid. These counter cations are immobilised to a greater or lesser extent by electrostatic forces. The net effect is a series of ionic layers very close to the clay surface. The system was very well documented by Winsauer and McCardell (1953).

Initially the electrical conductivity due to clay was proportionally added to that due to the water phase using simple equations of the form:

$$\frac{1}{R_z} = \frac{1-q}{R_w} + \frac{q}{R_{cl}}$$

where R_z is the resistivity of the 'clay slurry' in the pore, R_w is the resistivity of the water phase, and q is the fraction of the pore volume occupied by clay of resistivity R_{cl}, or

$$\frac{1}{R_0} = \frac{1-V_{cl}}{FR_w} + \frac{V_{cl}}{R_{cl}}$$

where R_0 is the rock resistivity at 100 % water saturation, F is the formation resistivity factor corresponding to the effective porosity ϕ_e, i.e. $F = a/\phi_e^m$, and V_{cl} is the fraction of the total rock volume occupied by clay of resistivity R_{cl}. More complex equations were suggested by Simandoux (1963) and by Poupon and Leveaux (1971) to relate the rock resistivity to S_w in the presence of clay, and these equations were, and still are, used extensively. These are of the form:

$$\frac{1}{\sqrt{R_T}} = \left[\frac{(V_{cl})^d}{\sqrt{R_{cl}}} + \frac{\phi^{m/2}}{\sqrt{(aR_w)}}\right] S_w^{n/2}$$

or alternatively

$$\frac{1}{R_T} = \frac{V_{cl}S_w}{R_{cl}} + \frac{\phi^m S_w^n}{aR_w}$$

where: n is the Archie resistivity index exponent; a and m are the constants in the Humble relationship $F = a/\phi^m$; $d = (1 - V_{cl})/2$ where V_{cl} is the volume fraction of clay; and R_T is the measured resistivity at a water saturation fraction of the pore space S_w.

Further advances were due to the recognition that the excess conductivity due to the presence of clay was a function of the water immobilised by the clay, i.e. the clay-bound water, CBW. This water surrounds the clay in equilibrium with the environmental electrolyte, and contains the counter cations which compensate the negative charges on the clay surface; it is anion free. The counter cations constitute the cation exchange capacity, CEC (in meq/g of dry rock) or Q_v (in meq/cm^3 total pore volume). According to Juhasz (1979), the clay hydration water, W_s (g/g dry rock), is proportional to CEC and inversely proportional to the salinity C_0 (meq/cm^3) of the equilibrium solution, which gave the relationship:

$$W_s = (0{\cdot}084C_0^{-1/2} + 0{\cdot}22) \quad \text{CEC}$$

from which he derived the equation:

$$\phi_e = \phi_T \left[1 - (0{\cdot}084C_0^{-1/2} + 0{\cdot}22)Q_v \frac{1}{\rho_{CBW}} \right]$$

where

$$\phi_e = \phi_T - \phi_{CBW}$$

Waxman and Smits (1968) and Waxman and Thomas (1974) defined the contribution of clay counter cations to the electrical conductance of a shaly formation to be a function of the concentration of the clay counter cations Q_v/S_{wt}, and B, their equivalent conductance [(mho/m)/(meq/cm^3)]. They also demonstrated that B was a function of salinity and temperature. The equation for S_{wt} is:

$$S_{wt} = \left[\frac{a\phi_T^{-m^*} R_{we}}{R_t} \right]^{1/n^*}$$

where

$$R_{we} = \frac{1}{C_{wt} + BQ_v/S_{wt}}$$

and m^* is the exponent in the relationship between F^*, the formation resistivity factor, and total porosity ϕ_T, under conditions such that F^* is independent of clay conductivity but includes the effect of the clay on pore tortuosity. An approximation for B was given by Juhasz:

$$B = \frac{1{\cdot}28 + 0{\cdot}255t - 0{\cdot}000\,405\,9t^2}{1 + R_w^{1{\cdot}23}(0{\cdot}045t - 0{\cdot}27)}$$

where t is temperature (in °C) and R_w is the equilibrium water resistivity (in ohm m).

CEC was originally measured on core and samples, but, subsequently, much of the literature has been devoted to methods of estimating Q_v and B.

A slightly different model was proposed by Clavier *et al.* (1977). The 'dual water' model of Clavier and his co-workers was developed to overcome two problems encountered in the use of the Waxman and Smits model. First, Q_v inferred from core conductivity measurements could be less than that indicated by CEC measurements. It was concluded that CEC not originating from clays could have somewhat different electrical properties from the clay CEC. The second problem is that shale-bound waters tend to have low salinity, almost independent of the surrounding waters. This is not implied by the Waxman–Smits model.

According to the dual water model the zone near to the clay surface, where Na^+ exceeds Cl^-, constitutes the 'diffuse layer'. Its thickness is characterised by X_d, which is given by:

$$X_d \text{ (in Ångströms)} = 30\sqrt{\frac{1}{n}}$$

where n is the salt concentration at 25 °C in the bulk water (moles/litre).

The Stern model accounts for the fact that the compensating Na^+ ions of the diffuse layer are kept away from the clay surface by the hydration water around the cation and the water absorbed on the clay surface. The inner boundary of the compensating Na^+ ions is the outer Helmholtz plane at a distance X_H from the clay surface, $X_H = 6{\cdot}18$ Å.

There is therefore a zone of salt exclusion around the clay particle which has a constant thickness X_H, when the bulk water salinity $[n]$ is greater than $[n_1]$ such that $X_H > X_d$. When the water salinity is less than $[n]$ the value giving $X_H > X_d$, the thickness of the zone of salt exclusion, is salinity dependent and equal to X_d. $[n_1] = 0{\cdot}236$ mols/litre NaCl. In order to compute the volume of the salt-free water the surface area of contact between the clays and the water is required. Experimental data have indicated that the specific surface area A_{sp} of clays is directly proportional to specific CEC (meq/g) irrespective of the type of clay. That is,

$$A_{sp} = V(\text{CEC})_{sp}$$

and the clay surface area per unit pore volume is

$$A_v = VQ_v$$

Developing the analysis, the relationship between the conductivity C_0 of the water-saturated shaly reservoir, and the conductivity C_w of the saturating water, was derived as:

$$C_0 = \frac{1}{F_0}[(1 - \alpha V_q Q_v)C_w + \beta Q_v]$$

where F_0 is the formation factor, β is the equivalent conductivity of the clay-bound water (salt-free zone) and $\alpha = 1$, for salinities $[n]$ greater than the limiting value $[n_1]$ or $\alpha = \sqrt{([n_1]/[n])}$ for salinities $[n] < [n_1]$.

In water-wet rocks, interfacial tension maintains hydrocarbons in the central part of the pore space where the 'far' or normal water is located. Assuming this water is displaced first, then at water saturations greater than the volume fraction of the total porosity occupied by clay bound ('near') water, the derived formation conductivity C_t is given by:

$$C_t = \frac{1}{F_0} S_{wt}^n \left[C_w + \alpha V_q \frac{Q_v}{S_{wt}} (C_{CBW} - C_w) \right]$$

where C_w the far water conductivity, C_{CBW} the clay-bound water conductivity and V_q are at formation temperature.

When all the 'far' water has been displaced, $S_{wt} \leq \alpha V_q Q_v$, the hydrocarbons may displace some of the hydration or absorbed water. (The counter ions remain to maintain electrical neutrality.) In this case $C_{we} = \beta Q_v / S_{wt}$ and

$$C_t = \frac{S_{wt}^n}{F_0} \frac{\beta Q_v}{S_{wt}} = \frac{S_{wt}^{n-1} B Q_v}{F_0}$$

Graphs of V_q, β and C_{CBW} $(= \beta / V_q)$ against temperature were given by Clavier.

The dual water model therefore postulates that: a shaly formation behaves as if it were clean, but contains a mixture of two waters; the formation or 'far' water is the normal reservoir water; the 'clay' or near water has a universal conductivity dependent only on temperature; when all of the far water has been removed from a clay by compaction the result is a perfect shale having

$$Q_{v_{sh}} = \frac{1}{\alpha V_q}$$

For Na^+ clays the compensating ion concentration reaches a maximum $Q_{v_{sh}} = 1/V_q$ depending on temperature and varying from 3·3 to 6·6 meq/cm^3. The effective clay water conductivity is $C_{CBW} = \beta / V_q =$ 6·8 mho/m at 25 °C and the Na^+ transport number $T_{Na}^{sh} = 1$, giving a straight, maximum value, shale base line on the SP curve.

Currently most serious workers use either the Waxman–Smits or the dual water model for detailed interpretation of shaly reservoirs. Each has its advantages and disadvantages, and investigations into the various problems continue.

2.3. Extension of Interpretation to Other Parameters of Interest

For many years the main application of well log data was to the more accurate definition of hydrocarbons in place and the emphasis was on the computation of porosity and hydrocarbon saturation. While this remains true, there has been a most gratifying increase in interest in the derivations of other characteristics concerning the reservoir at the exploration and development phases. Two major areas of interest have been in the identification of zones of anomalously high fluid pressure (one of the problems highlighted in UK North Sea development), and in the estimation of the mechanical strength of the reservoir rocks. The former certainly has led to an increasing concern in mud logging methods and measurements, cuttings analysis, and a more general integration of all data concerning downhole conditions. The latter, while in its infancy, has already suggested areas of common interest between engineering geology in the civil engineering field, and the petroleum geologist and petrophysicist.

The derivation of permeability from *in-situ* downhole measurements has always been a subject of considerable importance. This has led to a revival of interest in nuclear magnetic measurements, and in many factors affecting the transmission of acoustic waves through porous media, one of which is certainly the nature of the interfacial forces between the liquid and the solid, which are also forces involved in the hydraulic permeability of porous media.

Reservoir monitoring using well log and allied data has become an accepted routine. This resulted primarily from the development by the logging service companies of reliable through tubing thermal neutron die away measurements. These respond primarily to the presence of chlorine and hence to the amount of saline reservoir water present. The success obtained when high salinity water was present highlighted the requirement for practical methods responsive to the presence of low salinity water through tubing and casing, in order to monitor water movement with production. In parallel with this initial reservoir monitoring has been the application of multiple fluid pressure measurements in appraisal and development wells using repeat or multiple formation testers. Finally, the whole field of downhole production monitoring tools has undergone a consistent improvement.

The onset of the so-called energy crisis created an incentive for redoubling efforts on secondary and tertiary recovery of the hydrocarbons remaining in the reservoir towards the end of primary and secondary recovery. Similarly, there has been an increase of effort and investment in the 'more difficult' hydrocarbon accumulations such as the heavy oil and tar sands. Particular emphasis in this respect has been on fracture identification and location. In addition, the application of well logging to the coal and minerals (uranium) industries is of increasing importance. It was natural that in the early stages of work on these specialised applications, routine techniques were employed. However, the increase of interest in, and of the importance of much of the work, has renewed the effort on development of specialised techniques. These in turn can be used to assist in solving routine problems, and advance the whole of the science of formation evaluation.

3. RADIOACTIVE MEASUREMENTS

The specific requirements of measurements through tubing and casing in production wells, for exploration and production of uranium, and particular problems in the coal industry, such as the proportion of sulphur present in the coal, have all led to a broad expansion of knowledge of, and interest in radioactive measurements.

Downhole tools for measurement of natural γ-radiation have been available since the early 1950s. Natural γ-activity results from radioactive decay of the K-40 isotope of potassium, and from the uranium-238 and thorium-232 series. The latter two isotopes consist of a series of radioactive daughter elements each decaying at a specific rate, and each giving rise to specific radiation (not necessarily γ-photons). Naturally occurring potassium produces 3·4 γ-photons/s/g having an energy of 1·46 MeV excluding γ-photons having energies below 100 keV. Uranium in equilibrium with its daughter products produces $2{\cdot}8 \times 10^4$ γ-photons/s/g having a mean energy of 0·8 MeV. Similarly, thorium in equilibrium with its daughter products produces 1×10^4 γ-photons/s/g having a mean energy of 0·93 MeV.

Development of energy-sensitive γ-ray detectors made it possible to measure the energy of γ-photons impinging on the detector. Energy 'windows' can be set in the measurement system enabling a count rate to be made of γ-photons of a given energy band. This enables the identification and often quantitative estimation of the element producing the response. In practice the presence of thorium is identified by measuring the γ-photons having an energy of 2·62 MeV; these, in fact, emanate from the thorium daughter product thallium-208. Likewise, uranium is monitored by utilising the 1·76 MeV γ-photons emitted from the uranium daughter bismuth-214.

Quantitative assessment of radioactive elements present in the formation is quite a complex procedure, involving extremely careful calibration of the spectral γ-measuring system. Although neutron–γ and neutron–neutron porosity devices have been available since the late 1950s the interactions of neutrons with matter give rise to a host of interactions of possible value to formation evaluation. However these could not be properly exploited until the recent improvements in measurement technology and data processing. High energy or fast neutrons, in passing through matter, can collide with the various nuclei present and lose energy by elastic collision. They can also interact inelastically with the nuclei, giving up some or all of their energy, and leaving the target nucleus in an excited state. The excited nucleus will decay with the prompt emission of energy, often in the form of γ-photons. The energy of the γ-photons is specific to the target nucleus, and can hence be used both to identify it and to assess the quantity present. The fast neutrons can also enter specific nuclei which are transformed by the neutron activation, usually producing an active isotope which decays at a defined rate specific to the isotope and gives rise to radiation having energies specific to the isotope. Each element has a neutron scatter cross-section which determines the efficiency with which the element will scatter the neutrons. The elements also have activation cross-sections describing the probability of their entering into an activation type interaction. These cross-sections are measured in barns (1 barn $= 10^{-24}$ cm^2). The cross-sections are energy dependent, and are often zero below a certain neutron energy threshold, depending on binding energies within the target nucleus.

The amount of energy lost by a neutron undergoing elastic scatter is dependent on the mass of the target nucleus—the lighter the nucleus, the greater is the energy lost by the neutron. The loss of energy by elastic scatter continues until the neutron is in thermal equilibrium with its environment, when it has thermal energy (about 0·025 eV) and is defined as a thermal neutron. Thermal neutrons undergo thermal capture by nuclei with which they interact. Each nucleus has a specific thermal capture cross-section defining the probability of the occurrence of thermal neutron capture. The capturing nuclei are excited and lose energy promptly in the form of photons, which have energies specific to the nucleus involved.

The lightest element hydrogen absorbs the most energy when involved in an elastic collision with a neutron. Hence the distance that fast neutrons travel before thermalisation is reduced as the hydrogen content of the environment increases—this is the basis of neutron porosity logs which effectively measure the efficiency with which the formation slows down or moderates the fast neutrons emitted from the neutron source. The sources used continuously emit neutrons and hence the porosity-sensitive neutron tool is a steady-state tool, equilibrium between fast neutron emission by the source, thermalisation by the environment and thermal neutron capture in the environment is set up very quickly (a few milliseconds).

A major breakthrough resulted from the development of neutron generators which could be pulsed, producing controlled bursts of fast neutrons, and which could be operated downhole. Using these generators, a burst of fast neutrons is emitted by the source into the formation. The fast neutrons are moderated to give a thermal neutron flux in the vicinity of the source, which persists after the fast neutron burst. The thermal neutrons are captured by the various nuclei present in proportion to their thermal neutron capture cross-section and their relative abundance. Fortunately, one of the commonly occurring elements having a high thermal neutron capture cross-section is chlorine. Therefore, the presence of chlorine in reservoir water will cause an increase in the rate at which the thermal neutron flux decays. As a corollary, in a hydrocarbon-bearing zone the neutron flux will decay more slowly, the less the saturation of saline reservoir water. This is the basis of the thermal neutron die away tools which measure the rate of decay of the thermal neutron flux produced in the reservoir by a burst of fast neutrons. The basic equation for the rate of thermal flux decay is:

$$\log \frac{N_1}{N_2} = \Sigma V(t_2 - t_1)$$

where N_1 and N_2 are the thermal neutron fluxes present at times t_1 and t_2 (in seconds), V is the thermal neutron velocity (cm/s) and Σ is the macroscopic thermal neutron capture cross-section of the formation (Σ is the proportionate sum of the contribution of the individual components of the environment). For a reservoir rock having a porosity ϕ, the measured Σ is given by

$$\Sigma = (1 - \phi)\,\Sigma_{ma} + \phi\,\Sigma_f$$

where Σ_{ma} is the capture cross-section of the rock matrix usually about 5–10 capture units, and Σ_f is the capture cross-section of the pore-filling fluid. For oil, Σ_o is about 7 capture units and for reservoir water Σ_w depends on the salinity (chlorine content) and can be as high as 100 capture units for salinities of 220 000 ppm NaCl. (One capture unit = $10^{-3}\,cm^2$.) For a porosity ϕ and a water saturation fraction S_w of the pore space the measured Σ_m is given by:

$$\Sigma_m = (1 - \phi)\,\Sigma_{ma} + \phi[S_w\,\Sigma_w + (1 - S_w)\,\Sigma_o]$$

If a volume fraction of clay V_{cl} is present, having a capture cross-section Σ_{cl}, then

$$\Sigma_m = (1 - \phi - V_{cl})\,\Sigma_{ma} + V_{cl}\,\Sigma_{cl} + \phi[S_w\,\Sigma_w + (1 - S_w)\,\Sigma_o]$$

The value of the thermal neutron die away measurement is that it is little affected by the presence of casing and tubing. The measurement can therefore be applied during the course of depletion of a reservoir to monitor the rise of bottom water with production. This is achieved by logging at regular (6 monthly or yearly) intervals, often described as time-lapse logging. It has also been used in high permeability reservoirs (e.g. in the Middle East) to monitor the increase in the height of water coning round a production well with increase in offtake. The efficient use of thermal neutron die away techniques is limited to reservoirs where the water salinity is in excess of 50 000–60 000 ppm NaCl (depending on porosity).

Newer tools with more stable and more sensitive systems are reducing these salinity limits. However, to overcome the problem of monitoring water encroachment in low-salinity water environments interest was focused on spectral analysis of the γ-radiation emitted by the various nuclei interacting with the neutrons. This led to the development of the carbon/oxygen logging device. Both carbon and oxygen interact inelastically with the fast neutrons in the emitted neutron burst, giving rise to prompt γ-photons having a range of energy levels. The most important of these for carbon are at 3·41, 3·92 and 4·43 MeV, and for oxygen at 5·11, 5·62 and 6·13 MeV. (In each case, the two lower energy peaks separated by 0·51 MeV are functions of the crystal γ-ray detector used.) Energy gates within the tool measurement system are set to cover these two ranges. The count rate in the higher energy gate responds to changes in the oxygen content in the environment, while that in the lower gate responds to changes in the carbon content of the environment. If all conditions remained constant over the region of interest in the reservoir, the ratio of counts in the two gates would reflect changes in the oil (carbon)/water (oxygen) saturation of the rock. One immediate problem is that calcareous rocks contain carbon as carbonate in varying proportions, and most rocks contain oxygen. Fortunately both of the significant elements present in sedimentary rocks, silicon and calcium, give two sets of definable γ-radiation spectra as a result of interaction with neutrons. The first is as a result of inelastic scatter of fast neutrons, which obviously occurs during the neutron burst from the source. The second is as a result of thermal neutron capture, occurring for hundreds of microseconds after the source is switched off. The latter is used to measure the calcium/silicon ratio and hence allows a correction for the nature of the lithology. The lithology measurements are made by opening appropriate measurement gates in the quiescent period between the neutron bursts from the source.

There still remains a range of problems to be solved for the carbon/oxygen logging tool. Full energy spectra of the γ-photons produced during the neutron burst by inelastic scatter of fast neutrons will exhibit peaks from a whole range of elements. The spectra of the thermal neutron capture γ-photons produced in the quiescent period will similarly produce a multitude of energy peaks. As the thermal neutron capture regime can persist for several hundreds of microseconds, the resulting γ-photons can be present during subsequent neutron bursts. As mentioned earlier, thallium-activated sodium iodide crystal detectors, commonly used in downhole logging tools, give rise to a triad of peaks for the high energy γ-radiation absorbed within the crystal. The two lower energy 'escape peaks' are due to the formation of a positron–electron pair with the loss of one (0·51 MeV) or both (1·02 MeV) of the quanta produced when the positron is annihilated. To make matters even more complex the γ-photons emitted from the nuclei within the formation, at the energy specific to the nucleus, do not maintain their energy. The γ-photons lose energy by collision with electrons associated with the target nuclei, and hence can arrive at the tool detector with considerably degraded energy.

The spectra of γ-photon energies measured at the detector is therefore far from straightforward. Resolution of such spectra often necessitates complex spectrum stripping programs to be run on a computer in order to resolve the true intensity of a defined energy peak.

Interest in the various neutron interactions spreads well beyond the measurement of carbon/oxygen ratios in oil reservoirs. A large number of elements give rise to detectable radiation when subjected to neutron bombardment. In most cases of interest the induced radiation is in the form of γ-photons, although in the minerals field there has been considerable effort in the use of high energy (hard) x-rays. However, these are limited by their very much lower penetrating power. One special exception in the mineral field, is that of uranium which has a specific interaction with thermal neutrons producing fast neutrons as a result of fission in uranium-235. Thus, if uranium is present in a formation subjected to a burst of fast neutrons, the fast neutrons are thermalised; the thermal neutrons induce fission in the uranium-235, resulting in further fast neutrons. These in turn are moderated, but are detected as epithermal neutrons, having energies just above the thermal neutrons. Increase in this epithermal neutron flux is proportional to uranium concentration.

Solid-state detectors, which have higher resolution to γ-ray energy levels than crystal detectors, have been available for some time. However they need to be operated at low temperatures. Prototype logging tools using cryostatically cooled solid-state detectors have been tested. It is to be expected that development of more efficient γ-radiation detectors which do not need the present high degree of cooling will continue. In the radioactive measurement field, therefore, there are exciting future possibilities for formation evaluation.

Modern computer technology has made it possible to carry out complicated computations of the flux distributions set up by radioactive sources in borehole environments. Without the modern computer these are laborious and very time consuming to the point of being impractical. Such studies have led to better understanding of, and resultant improvement to, the conventional radioactive tool.

The early neutron porosity tool had an isotope neutron source (radium/beryllium) with a single detector. This latter was either a thermal neutron detector or a γ-radiation detector responding to the prompt γ-radiation flux due to thermal neutron capture (proportional to the thermal neutron flux). In both cases, the measured flux was clearly affected by the thermal neutron capture cross-section of the formation; for example, the flux at the detector is lower for salt water filled porosity than for fresh water or oil filled. The early solution was to use an epithermal neutron detector skid mounted with the source, and with the skid pressed against the borehole wall. Analysis of the total neutron flux distribution showed that the ratio of thermal neutron fluxes at two points remote from the source, was related to the epithermal neutron flux and was essentially independent of capture effects. Increase in source detector distance also reduced the effect of the borehole on the response, and gave an increase in effective depth of investigation of the device. The basic investigation led to the development of compensated neutron tools using two detectors spaced as far from the source as possible. This in turn requires a high source strength (i.e. a high neutron output) to reduce statistical counting errors. The practical limit is therefore constrained by the hazard to personnel handling high yield sources. This has been ameliorated by the discovery in 1950 of the artificial isotope californium-245, which has a high neutron yield per gram. Californium neutron sources are being increasingly utilised in the field of neutron logging.

Modern density tools have undergone a similar development to reduce unwanted effects on the response. The effect of the borehole was initially reduced by mounting the γ-radiation source and detector in a shielded skid pressed against the borehole wall. This left the effects due to the presence of a mud cake; these were effectively overcome by using two detectors and relating the ratio of their responses to formation density. The effects due to the presence of a mud cake on this device are relatively easy to correct. The density tool measures the overall attenuation of the γ-radiation flux by the medium between the source and detector. The relevant methods of interaction between γ-rays and matter are ion-pair production, Compton scattering and photoelectric absorption. The absorption coefficient for pair production has a threshold at 1·02 MeV, and for media of interest, pair production is very small below about 10 MeV. The lack of suitable high energy γ-radiation sources and the fact that the cross-section is Z-dependent, relating the response to chemistry of the medium, rules out the pair production interaction for density logging purposes. For γ-radiation energies between 1·02 MeV and 0·1 MeV the only interaction of interest (excluding the heaviest elements) is Compton scattering. This is the controlling mechanism for conventional density logging. In this regime the γ-flux N_x, crossing unit cross-sectional area at a distance x cm from and perpendicular to, a source emitting N_0 γ-photons/s is given by:

$$N_x = \frac{\exp(-\mu x)}{4\pi x^2} N_0$$

where μ is the Compton scattering cross-section.

The γ-photons are scattered by the electrons associated with the atomic nuclei, and the scatter cross-section σ_s of an electron, is a function of the γ-radiation energy.

The number of the scatters per cm^3 is therefore the product of σ_s and the number of electrons per cm^3. There are $6{\cdot}023 \times 10^{23}$ atoms per gram atom (Avogadro's constant). The number of atoms per gram is therefore $(6{\cdot}023 \times 10^{23})/A$, where A is the atomic weight. The number of electrons per atom is the atomic number Z; hence the number of electrons per gram is

$$\frac{(6{\cdot}023 \times 10^{23})Z}{A}$$

For mixtures and compounds containing different atoms:

$$\text{Electrons per cm}^3 = 6{\cdot}023 \times 10^{23} \left(\sum \frac{Z_i C_i}{A_i}\right) \rho$$

where $\rho =$ bulk density in g/cm^3.

Hence

$$\mu = (6{\cdot}023 \times 10^{23})\sigma_s \rho \frac{Z}{A}$$

where

$$\frac{Z}{A} = \sum \frac{Z_i C_i}{A_i}$$

and $\sigma_s = 0{\cdot}2/10^{24}\,\mathrm{cm}^2$ for the energies in question, then

$$\mu = 0{\cdot}12\rho \frac{Z}{A}$$

and

$$N_x = \frac{N_0}{4\Pi x^2} \exp\left(-0{\cdot}12\rho \frac{Z}{A} x\right)$$

i.e.

$$N_x = K_1 \exp\left(-0{\cdot}12\rho \frac{Z}{A} x\right)$$

or

$$\log N_x = K_1 - K_2 \frac{\rho Z}{A}$$

For most sedimentary rocks Z/A is very close to 0·5. However a lithology correction is required for accurate work as is also a correction for water and hydrocarbon in the pore space.

$$Z/A\,(H_2O) = 0{\cdot}555 \qquad \text{and} \qquad Z/A\,(CH_2)_n = 0{\cdot}570$$

Original density tools had the γ-radiation source (cobalt-60 or caesium-137) and the detector shielded and collimated in a skid which was spring loaded against the borehole wall. The major error was caused by the presence of mud cake between the skid unit and the formation of interest. Modern tools use a two-detector system which allows much more accurate assessment of the mud cake.

The correction is automatically applied during logging with a record on the log (and tape) of the amount of correction applied. The third method of interaction of γ-photons with matter is by photoelectric absorption. The photoelectric absorption coefficient is very small for γ-radiation energies above 0·1 MeV but increases sharply as the γ-radiation energy is reduced. It also varies approximately proportionately to $Z^{4\cdot6}$, and is therefore extremely sensitive to lithology. The most recent γ–γ tool development is the lithology tool which measures the index of the photoelectric cross-section P_E, where

$$P_E = \left[\frac{Z}{10}\right]^{3\cdot6}$$

The measurement of the P_E index is combined with a conventional density recording (Compton scatter). As with conventional density the lithology tool responds to the electron density ρ [$=6\cdot02 \times 10^{23}(Z/A)$] which is related to an electron density index ρ_e where ρ_e is defined as

$$\rho_e = \frac{2}{6\cdot023 \times 10^{23}} \times \text{electron density}$$

(which for $Z/A = 0\cdot5$ gives $\rho_e = \rho$).

The effective photoelectric absorption cross-section index U, per unit volume, has the units of barns/unit volume and is given by:

$$U = P_E \rho_e$$

The cross-section parameter U is additive for the components of a formation as is the density. Thus, the two measurements of photoelectric absorption cross-section index, and bulk density, give two similar equations, viz:

$$\rho_b = (1-\phi)\rho_{ma}^{+\phi\rho_f}$$
$$U = (1-\phi)U_{ma}^{+\phi U_f}$$

The value of the P_E measurement is that it is primarily influenced by the nature of the rock matrix, and only secondarily by the porosity. It can be combined with the conventional density and neutron logs for resolving multiple rock matrix problems. The presence of barite is, however, detrimental to the measurement, due to the high value of 267 for its P_E, compared with 1·81, 5·08 and 3·14 for quartz, calcite and dolomite respectively.

The photoelectric responsive γ–γ log is very new, and untried. It should however prove to be an important aid to formation evaluation.

In the last decade or so, radioactive logging has developed into one of the most useful additions to basic formation evaluation. It is believed that the potential for its future development is extremely large.

4. ACOUSTIC MEASUREMENTS

The sonic log, measuring the travel time of a compressional acoustic wave, has been in continuous use since about 1951. The popularity of the tool was due to the simple time-average equation postulated by Wyllie and his co-workers in 1956. The linear response, relating total response to the proportionate responses of the component parts of the porous medium, provided a simple method of relating travel time ΔT to porosity, i.e.

$$\Delta T = \phi \Delta T_{\mathrm{f}} + (1 - \phi) \Delta T_{\mathrm{ma}}$$

where ΔT_{ma} and ΔT_{f} are the acoustic compressional wave travel times in the rock matrix and pore-filling fluid, respectively, and ΔT is the travel time of the first compressional wave arrival. Unfortunately, the limitations on the conditions under which the linear response relationship was originally postulated, have often been forgotten and ignored.

Early tools used one transmitter and two receivers, measuring the difference in the travel time from transmitter to each receiver, thus cancelling or minimising the travel time in the borehole fluid from the transmitter to the borehole wall and from the borehole wall to the receivers. The acoustic wave is critically refracted at the borehole wall, and hence the depth of investigation of the device is small. The conventional modern tools achieve better compensation for borehole effects by using two transmitters with two or more receivers in an array between them. The measured formation travel time is the mean of those measured by two mirror image single transmitter, two-receiver arrays. The object of the borehole compensated sonic tool is to minimise errors due to unequal travel time in the borehole fluid paths caused by caving and tilt of the tool axis relative to the borehole axis.

In the late 1950s/early 1960s the measurement of the attenuation of the acoustic compressional wave led to the development of the cement bond log, CBL. Early tools measured the amplitude of the first compressional wave (or P-wave) arrival in the cased hole. Good bonding of cement to the casing and to the formation causes a high dissipation of energy to the formation, resulting in a low amplitude-received signal. At the other extreme, completely uncemented pipe transmits the acoustic energy at the pipe velocity relatively efficiently, resulting in a good amplitude signal. The range of possibilities between these two conditions caused considerable difficulties in CBL interpretation. One in particular was the presence of a micro-annulus between the casing and cement; this disappeared if the casing was internally pressured. Unfortunately, such a micro-annulus, when it exists, must tend to open under producing conditions when the pressure gradient is from formation to wellbore. (It cannot of course be remedied by squeeze cementing as the immediate effect of the squeeze is to close the annulus.) The problems with interpretation of CBL results focused interest on additional information in the complete acoustic signal at the receiver. Modern CBL tools present the total signal either as a full wave recording or in a variable density display where the peaks and valleys of the waveform are converted to a black/white presentation, in which the density is a function of the signal amplitude. These displays allow both the casing signals and the later formation signals (when present) to be examined.

The physics of the propagation of acoustic waves in the earth has of course been of interest to geophysicists and seismologists for many years. The development of acoustic logging methods focused attention on the more detailed physics of acoustic wave propagation in porous media, and in the borehole environment. The initial basic work can be credited to Biot in the early 1950s. Since that time many workers have contributed to the basic knowledge of the subject. Differences of opinion still exist concerning the nature of the various arrivals which can be identified from a full wave recording of the received signal. In acoustic logging, as in the other forms of logging, a number of basic problems still remain to be resolved. The major difficulty is the quantity and complexity of the information contained in the waveform records. Recording and analysis of the data have been considerably helped by the development of modern data handling and data processing methods.

Two types of waves propagate from the borehole. First, there are two body waves travelling at velocities dependent on the elastic constants of the medium and critically refracted at the borehole wall. The first is the compressional or P-wave, and the second is the shear, or S-wave. Since fluid will not support the transmission of an S-wave, the S-wave travelling in the formation is formed by mode transfer of the P-wave, at the borehole/formation interface. The sequence therefore is P-wave from transmitter to borehole wall, mode transfer to S-wave, S-wave travel in the formation, mode transfer to P-wave at the borehole wall and travel to receiver as a P-wave.

The P-wave velocity V_P is given by:

$$V_P^2 = \frac{1}{\rho_b}\left[K + \tfrac{4}{3}\mu\right] \qquad (\text{cm/s})^2$$

and the *S*-wave velocity, V_S, is given by:

$$V_S^2 = \frac{\mu}{\rho_b}$$

where K is the bulk elastic modulus, μ the shear modulus, and ρ_b the density of the medium. The shear wave will only appear if V_S in the formation is greater than V_P in the borehole fluid. In most formations V_P is from 1·6 to 2·0 times greater than V_S.

The second type of wave propagated in the borehole is the guided wave. The first to arrive is the Rayleigh wave (or pseudo-Rayleigh wave according to some workers). It is highly dispersive and appears as a long ringing train immediately following the *S*-wave arrival. The second guided wave is the Stoneley wave, which is slightly dispersive, and appears as a lower frequency pulse in the middle of the Rayleigh wave. The frequency of the Rayleigh wave before the Stoneley wave is recognisably higher than that following.

Various improved algorithms have been proposed to improve the accuracy of deriving porosity from measurements of V_P. More fundamental relationships based on the physics of the wave propagation are required. It would appear probable that these will initially relate to elastic constants, and use the P- and S-wave measured velocities to measure the required constants. Of course the *in-situ* elastic constants of rocks are of value in their own right to estimate drilling characteristics, etc., and to the engineering geologist.

Measurement of the attenuation of acoustic waves in the borehole is required for correction of seismic reflection data, for generation of synthetic seismogram sections and for lithological correlation. Recent work indicates that attenuation is related to the saturation state and to the nature of the saturating fluids in the porous media. It increases with decreasing effective pressure and with increase in permeability.

Increase in source–detector spacing achieves two benefits. First, the longer spacing increases the effective depth of investigation into the formation. Secondly, the time separation between events resulting from different travel times is increased, thereby (in principle) improving the resolution of the events. However, the increase in travel path decreases the signal amplitude, reducing the signal/noise ratio at the receiver. The situation can be improved by using a high power in the transmitter and higher sensitivity receivers. Recent developments have made this possible and long-spacing sonic tools, having multiple transducers and spacings of over 10 m, have been used experimentally. Commercially available long-spaced sonic tools are in service having spacings of the order of 10 ft. The received wave trains are digitally recorded, and can be subjected to modern mathematical filtration and signal-enhancement techniques. Analysis of the characteristics of the various types of wave propagated in the borehole can then be made.

One rather different type of acoustic logging tool, called the borehole televiewer, was introduced in the late 1960s/early 1970s. This comprised a small piezoelectric crystal ($\frac{1}{2}$ in diameter) which was pulsed at 1500 pulses per second in the thickness mode, giving a dominant frequency of 1·3 MHz. The emitted acoustic impulse was reflected back to the transducer which was switched to operate as a receiver. The transducer had a low Q to ensure that reverberation ceased before the received pulse. The transducer assembly was rotated at 3 rev./s and the systems triggered at magnetic north, producing a scan of the borehole having 485 data points every $\frac{1}{3}$ s. The record produced is an oriented display, intensity-modulated according to the amplitude of the received signal. The borehole logging speed was 300 ft/h so that each scan corresponded to $\frac{1}{3}$ in of borehole. The record represents the acoustic energy reflected into the receiver by the wellbore, which depends on rock reflectance, wall rugosity, wellbore fluid attenuation and hole geometry. The BHTV had an internal magnet to provide a synchronisation signal when used in casing.

There has been a steady increase in understanding of the mechanisms of acoustic wave transmission in the borehole environment in the last decade. Development of new acoustic logging methods and interpretation techniques must surely follow. This will certainly involve: better definition of mechanical strength properties of the rocks; more detailed examination of rock matrix properties for lithology resolution; additional information on degree of fluid saturation and saturating fluid type; better definition of porosity, possibly related to elastic constants, compressibility, etc.; and also permeability-related data.

5. FRACTURE IDENTIFICATION AND CHARACTERISATION

There has been strong emphasis over the past decade on methods of identifying fractures and, also, of characterising them. This has resulted partly from the interest of the petroleum industry in enhanced recovery. Other disciplines involved in radioactive waste disposal, geothermal projects, water resources and earthquake predictions, also have an increasing interest in accurate definition of porosity, permeability and engineering properties of fractured rock.

Most conventional logs give a predictable quantitative response to the presence of fractures and fracture zones. However, similar responses can be obtained from a range of conditions other than fractures. Any logging programme aimed at identifying and characterising fractures must therefore use a selected range of tools.

Open fractures intersecting a borehole will be flushed by drilling mud, and thus become paths of relatively high electrical conductivity. Given sufficient contrast between the conductivity of the drilling mud and that of the interstitial fluid (e.g. in hydrocarbon-bearing zones) fractures will cause an anomalous differential between the shallow and deeper investigative resistivity tools. Similarly, there will be anomalies in the measured formation resistivity factor–porosity relationship, which specifically relates to rocks having intergranular porosity.

Porosities computed from the density and neutron logs are total porosity values which include fracture porosity. On the other hand acoustic travel time tends to relate to the intergranular porosity. A low anomaly between sonic-derived and the density- or neutron-derived porosity can indicate the presence of vugular or fissure porosity. In addition the density correction curve recorded on the density log reflects the presence of rugosity, mud cake or fluid between the skid and the borehole wall. An increase in the correction applied together with negligible increase in porosity will often be noted opposite fracture zones. Increase in borehole wall rugosity due to fracturing can also affect the caliper curve especially if this is of the single arm variety.

Vertical or near-vertical fractures will be indicated on the 4 arm dipmeter curves by affecting opposite pairs of the arms. The excess conductivity due to the fracture will be shown on two arms 180° apart and not on the other pair. This can be identified in the processing of the dipmeter curves. The response is enhanced in the Schlumberger Fracture Identification Log, by running the tool at current saturation so that the relatively highly conductive fractures 'stand out' on the record. Non-vertical fractures can also be identified in the processing of dipmeter logs in that the fracture will tend to produce an anomaly in the general lithological dip pattern in the vicinity.

The presence of fractures has a strong effect in decreasing the amplitude of the acoustic compressional and shear (P and S) waves. This is also true for the Stoneley-type tube wave and other boundary waves. The theory of acoustic wave propagation in borehole geometry is still the subject of considerable work. However, the detailed study of full acoustic wave recording made with conventional sonic tools, and the development of four-pad sonic tools measuring acoustic characteristics from pad to pad round the borehole are all adding to the data available for study.

Borehole cameras and television recordings are extensively used in the water resources industry to examine fractures, etc., at the face of the wellbore. These are limited to application where the wellbore fluid is clear and transparent. An acoustic record similar to a television picture is provided by the borehole televiewer. This is finding an increasing application in the field of fracture location.

6. ELECTROMAGNETIC AND NUCLEAR MAGNETIC MEASUREMENTS

By the beginning of the 1970s all of the basic electromagnetic (resistivity) tools existed. The laterolog, induction and various microresistivity tools had been well tried, and were in regular commercial use. All of these processed in-phase electric current and potential measurements to provide resistivity or conductivity information. Increase in operational costs, particularly the escalation of rig costs offshore, led to the positive aim of reducing the rig time occupied by logging operators. The result was a series of combination tools, each producing a number of separate logs in one run in the borehole.

The laterolog 3 (guard electrode) and the laterolog 7 were succeeded by the laterolog 8 and laterolog 9 giving shallow and deep radial investigations. In 1970 the dual laterolog having nine electrodes was introduced. This gave a deep investigation curve LLD and a shallow investigation curve LLS, and could incorporate a γ-ray unit as well. Later versions included a shallow investigation micro spherically focused log, and an SP curve could be run on the cable bridle electrode. The dual laterolog–Rxo tool of Schlumberger uses different frequencies for the logging currents—35 Hz for the LLD, 280 Hz for the LLS and 1 kHz for the micro SFL. This tool was the first to enable three focused resistivity curves to be recorded in one run.

Similarly, the deep and medium induction logs were combined together with a shallow investigation laterolog 8. More recently the deep induction tool has been combined with a shallow investigation spherically focused device, and two curves run in conjunction with a borehole compensated sonic device. The laterolog 8 and spherically focused device operate at a frequency of 1 kHz and the induction device at 20 kHz. The various logging service companies have different combination logs under their own trade names.

Nuclear magnetic logging (NML) was introduced in 1960, essentially as a permeability-related measurement. It did not have a great success for several reasons, one of which was its shallow investigation (a few inches). However, the upsurge of interest in methods of secondary and tertiary recovery required accurate estimates of the residual oil remaining in the reservoir after primary depletion. In addition, such measurements were required in field experiments set up to optimise secondary and tertiary recovery techniques. This has led to an upsurge of interest in nuclear magnetic logging, which makes use of the fact that the hydrogen nucleus (proton) behaves as if it were a combined magnet and gyroscope. Oil and water contain approximately equal numbers of protons per unit volume; these are randomly oriented in the absence of a magnetic field. The application of a magnetic field tends to align a preponderance of the nuclei against the agitation due to thermal motion. The degree of polarisation is proportional to the strength of the applied field and inversely proportional to the absolute temperature. If, after polarisation is established, a new field is applied in a different direction this will exert a torque on the nuclei, which then precess due to the gyroscopic action. The frequency of precession is 4528 times the magnetic field strength (in gauss).

Modern nuclear magnetic logging devices use a long coil through which an electric current is passed to create a magnetic field in the formation, and polarisation of the protons present is built up. After a polarising time, T_p, the field is switched off leaving the protons under the action of the earth's magnetic field. The component of polarisation perpendicular to the earth's magnetic field precesses round the field, generating a current. The current in turn induces a voltage in the coil; this voltage is amplified and sent to the surface. (There is about 20 ms dead time in switching the coil function.) The signal is sinusoidal with an amplitude decay which is approximately exponential. The relaxation time is T_2, the transverse or spin–spin relaxation time. (T_1, the relaxation time for the component of polarisation parallel to the earth's magnetic field, is known as the longitudinal or spin-lattice relaxation time.)

While signals from a small sample will decay with time constant T_2, the time constant for the whole sample T_2^1 is less than T_2 due to inhomogeneities in the magnetic field. The nuclei in different parts of the sample also precess at slightly different frequencies and hence very quickly get out of phase with each other.

Relaxation time for water and oil by themselves vary approximately inversely with the viscosity. In a porous rock the solid surfaces act to accelerate the relaxation, giving much of the value to the NML measurement. In shale the specific surface area is sufficiently high to give relaxation times too short to be seen (of the order of a millisecond). NML responds to the fluid not affected by the pore surfaces, i.e. the 'free' fluid. In a water-wet rock the 'non-free' fluid is water. Magnetic formations cause very rapid signal decay. Hence paramagnetic ions added to the mud filtrate cause rapid signal decay from the borehole fluid, and from the water in the invaded zone. Therefore the measured signal is from the residual oil in the invaded zone. Manganese salts insoluble in oil are used for this purpose, usually with a chelating agent to prevent precipitation by alkaline mud, and/or removal in the mud cake or formation by cation exchange. Modern tools measure T_1 by comparing signals obtained using different polarising times T_p. T_2^1 is measured and used to extrapolate the waveform amplitude back to the beginning of precession to give the initial signal amplitude S_0 which is proportional to the number of protons per unit volume. This gives the free fluid index FFI. (The free fluid index unit is equal to 1% of the sample volume occupied by free fluid.) The signal frequency is also measured in order to give an accurate check on the earth's magnetic field.

Under normal reservoir conditions, the difference between total fluid and free fluid is the irreducible water saturation $S_{w_{irr}}$ which many workers have related empirically to permeability. If the invaded zone water is doped with paramagnetic ions, then the free fluid is the oil remaining in the reservoir, i.e.

$$\text{FFI} = \phi \times \text{Residual hydrocarbon} \times \text{Volume ration of hydrogen nuclei in the oil}$$

It is believed that NML will gain greater application in the future, which in turn will lead to further tool development. This will possibly follow the usual pattern of using multiple transducers (coils) in place of the single transducer which is currently used.

As mentioned earlier the conventional formation resistivity and conductivity responsive tools measure in-phase currents and voltages. Theoretically, direct current should be used. Indeed, the early discrete electrode resistivity measurements used pulsated direct current, and then developed into using relatively low frequency alternating current where the quadrature component was not important. Development of the induction log which has a frequency of a few kilocycles highlighted the fact that the quadrature component became very significant as the frequency of the logging current increased. This was known from other disciplines, work on cores and theoretical analysis of induced potential measurements used in the minerals industries. Dielectric constant measurement of reservoir rocks was of considerable interest, as the dielectric constant of water is significantly greater than that of oil, gas, and the basic sedimentary rocks, being 80, 2·2, 1·0 and 4–8, respectively (shale is about 13–15). However, measurement of the dielectric constant of the heterogeneous mixtures making up a reservoir rock gave values considerably higher than simple theory would indicate. Many organisations have been working on this problem for at least two decades, but many of the results have not been published at all, and some have only been published well after the work concluded.

When an electrical field is applied to a homogeneous material various mechanisms control the resulting effects. If the material contains free-moving ions or electrons, there is charge transport towards the electrodes applying the field giving the familiar electrical current and giving rise to conductivity losses. If no free charge movement is possible (as in insulators) there is a local shift of the centres of positive and negative charge, known as induced polarisation. This effect produces dielectric constant values in excess of unity (that of a vacuum). The shift in the centre of charge is very small and occurs almost instantaneously on the application of the field. When the material has dipolar molecules (e.g. water), the charge distribution is asymmetric. This causes some of the (originally) randomly oriented molecules to attain some degree of orientation in relation to the applied field. Due to the relatively large size of the molecules concerned, the orientation and relaxation as the field is applied and removed takes a finite time, which is defined by the relaxation time τ. If an alternating field is applied at a sufficiently high frequency, the dipoles cannot orientate quickly enough, and this effect no longer contributes to the total polarisation. A dispersion of the dipolar effect takes place about a frequency f governed by $2\pi f\tau = 1{\cdot}0$. For simple mixtures of components, phenomena such as interface charge effect complicate the simple theory.

The complex dielectric permittivity of the formation is ε given by $\varepsilon = \varepsilon' - j\varepsilon''$, where ε' is the real part and ε'' is the imaginary part accounting for the various loss mechanisms. The Schlumberger electromagnetic propagation tool (EPT) now available in the field, measures the propagation travel time t_{pe} and attenuation of a 1·1 GHz ($1{\cdot}1 \times 10^9$ Hz) electromagnetic wave travelling through the formation. These are related to ε' and ε'' by the equations

$$\varepsilon' = \frac{(t_{pe})^2}{\mu_0} - \frac{\alpha}{\omega^2 \mu_0}$$

and

$$\varepsilon'' = \frac{2\alpha t_{pe}}{\omega \mu_0}$$

where ω is the angular frequency, μ_0 is the magnetic permeability and α is the attenuation.

The tool is essentially salinity-independent as a result of its high frequency but only has a small investigation depth. The antenna system is pad mounted against the borehole wall and uses four antennae operating in a borehole compensating mode. It is not affected by thin mud cakes ($<\frac{3}{8}$ in) but is affected by borehole wall rugosity. It cannot be operated with oil or air in the borehole due to their low dielectric constants (and hence short propagation times). The tool can be run in combination with other telemetry compatible tools, in particular compensated neutron and density logs. Essentially, the tool responds to the bulk volume of water in the formation, and comparison of this with the total porosity provides the water saturation of the flushed zone. Published results using the EPT are very promising, and of course other companies are working on similar devices.

The major points of interest to the well logging discipline in the measurement of ε' is that ε' for a water-saturated rock is porosity-dependent and decreases with increase in frequency until about 250 MHz, when it has a value dependent on porosity, i.e. water content. Below 250 MHz, ε' increases with increase in salinity and the variation with porosity becomes somewhat unpredictable. There is clearly more work required to define the cause of some of the effects noted using frequencies above those used for conventional logging. Some of these effects may well be grain and pore geometry related, and hence could be valuable in a proper definition of the reservoir rock. However, the reduction in depth of investigation with increase in frequency will somewhat limit the practical application. In applications to secondary recovery, and exploitation of heavy oil and tar sands, a new generation of tools using VHF electromagnetic waves could prove invaluable.

The minerals industry has an interest in measuring magnetic susceptibility. In the past this has not proved practical, due largely to the generation of noise within the measurement system. However, it would appear that improvements in technology are remedying this problem and reliable tools should soon be available. Such tools could prove valuable in the petroleum industry in the cases where concentrations of ferromagnetic ores are high enough to give problems with conventional resistivity tools. Fortunately these cases are extremely rare.

Finally one of the areas in the field of electromagnetic wave measurement, where very significant improvements have steadily continued is in the measurement of stratigraphical dip. The basis of these improvements has been the advance in instrumental technology, data transmission and data processing. In the last decade or so, the borehole dipmeter has progressed from a tool which gave somewhat questionable results, to one which gives reliable data which can be cross-checked, and can be computer processed to give a wealth of useful data including identification and location of fractures.

7. MEASUREMENTS WHILE DRILLING

There are considerable technical incentives for measuring downhole data and telemetering them to the surface during drilling. In the mid 1960s the initial interest centred on logging while drilling. The priorities have changed since that time partly due to the changing economic climate, but largely due to users' opinions, as the possibility became more practical.

The technical advantage of logging while drilling is centred on obtaining formation data at an early stage of the filtrate penetration into permeable horizons. The practical economic advantage would be to ensure that some formation data are obtained in the event of losing the hole. It is unlikely in the foreseeable future to envisage that the ability to obtain sufficient data concerning the formation while drilling would enable one to dispense with conventional logging at a later date. This is more likely to occur if sophisticated methods of logging after casing are developed to give sufficient data for formation evaluation. This would reduce the amount of data required at the open hole phase, and hence reduce costly rig time, which is currently taken up with logging.

The present priorities for measurements while drilling are:

(a) Well safety—well control data.
(b) Directional optimisation to improve efficiency.
(c) Directional survey and control.
(d) Formation evaluation.

The development of sensors required for the basic measurements does not appear to present a major problem. Most real effort has been on the development of a satisfactory practical data telemetry system. This will control the accuracy and quantity of data which can be usefully acquired.

The incentives for developing real time measurement methods are very great, and a number of companies are involved in such work, including the major logging services. Hence the status of the work changes almost daily.

Basic telemetry systems use the mud flush as a 'prime mover'. The mud pressure pulse systems use a valve with its control system to modulate the resistance to flow in the mud system, creating pressure pulses which can be monitored at the surface. The pressure pulse travels at the velocity of sound in the mud (about 5000 ft/s). However, the data transmission rate is limited by pulse spreading and the limitation of the rate of modulation.

A second system, rather similar to an air siren, was introduced by Mobil; this uses a slotted rotating disc to generate a mud pulse carrier frequency in the range 100–300 Hz. The carrier was frequency-modulated to transmit the downhole data. At least one commercial company is using this principle in its development.

Several companies active in the field are working on electrical data transmission systems. Some of these require special drill pipe which includes the electrical cable system, and which requires special connectors at the drill pipes. Others are developing systems using continuous electrical cable lowered with the drill pipe. Excess cable is stored on a take up assembly inside the drill string. Companies involved in the development of electrodrills, which power from the surface, are also including conductors for data telemetry within the power cable package.

Transmission of electromagnetic signals either through the earth or up the drill pipe from downhole has also been studied. The increase in attenuation with frequency of electromagnetic waves passing through the earth limits their use to low frequencies. Telluric currents are also present at low frequency and this causes severe signal/noise problem with electromagnetic wave data transmission through the earth. Electromagnetic signals in the drill pipe are also subject to severe attenuation problems.

Acoustic transmission data systems operating under borehole conditions have also been the subject of considerable research, but at the present time there appears to be little, if any, published information on these.

Mud pressure pulse systems for transmitting data from the drilling bit to the surface are the most likely to be first in real commercial use.

The incentives for successful development are high. Success is therefore to be expected.

8. OTHER DEVELOPMENTS

Various advances in the basic wireline logging measurements have been discussed in some detail. However, there have been consistent advances in all aspects of the work. In the interpretation aspects, the application of modern data transmission and data processing methods has virtually revolutionised the utilisation of dipmeters. The utilisation of log data to detect overpressured formations is now a standard procedure. This was given considerable impetus during development of various UK North Sea fields.

Production tools have been used in combination and the running of production logs has, together with neutron die away measurement, become a standard post-completion procedure for production and reservoir monitoring. In parallel with this, the development of wireline formation testers which allow pressure to be measured at as many stations in the borehole as required has been enthusiastically received. These testers operate in open hole conditions but have been found to provide valuable information on reservoir depletion when run in development wells, drilled during the course of production.

Interest in fracture location and characterisation has renewed interest in the borehole televiewer. Interest in reef developments, in tight fractured gas-bearing carbonates, and in various geothermal research appears to be bringing renewed interest in borehole gravimeter measurements.

9. CONCLUSIONS

The interest in well logging applied to formation evaluation began with the petroleum industry and spread to the coal industry at quite an early stage. Over the recent past, the interest has spread to the minerals industry, to engineering geology, and to geothermal projects. Workers in all of these fields are routinely using wireline logs. More importantly, their own specialised requirements are leading to the development of specialised measurement systems and interpretation methods.

The broadening of interest, together with the increasing use of modern instrumentation technology and modern data handling and processing methods, have brought about a tremendous increase in the rate of development and the rate of advance in knowledge. This is certain to continue well into the next decade.

REFERENCES AND BIBLIOGRAPHY

(a) Use of Computers

BATEMAN, R. M. and KONEN, C. E. (1977) Well site log analysis and the programmable pocket calculator. *Trans. SPWLA 18th Ann. Symp.*, Paper B.

BEST, D. L. *et al.* (1978) A computer process well site computation. *Trans. SPWLA 19th Ann. Symp.*, Paper Z.

BUCKNER, G. O. (1964) Field digital recording of well log data. *Trans. SPWLA 5th Ann. Symp.*, Paper H.

BURKE, J. *et al.* (1967) Computer processing of log data enables better production in Chaveroo Field. *J. Pet. Tech.*, **19** (July), 889.

DEVRIES *et al.* (1976) The cased hole exploration service epilog. *Trans. SPWLA 4th European Symp.*, Paper D.

EATON, F. M. *et al.* (1976) The cyber service unit—an integrated logging system. *51st Ann. Fall. Tech. Conf. SPE of AIME*, Paper SPE 6158.

GILCHRIST, W. A. J. (1981) A sensitivity study of Schlumberger's cyberlook computation and its comparison to other evaluation methods. *Trans. SPWLA 22nd Ann. Symp.*, Paper F.

HOWELLS, A. P. S. and SCHOONOVER, L. (1977) Well log data by satellite. *Trans. SPWLA 18th Ann. Symp.*, Paper BB.

KRYGOWSKI, D. A. *et al.* (1980) Wellsite log analysis using a desk top computer. *Trans. SPWLA 21st Ann. Symp.*, Paper Y.

MAYER, C. and SIBBIT, A. (1980) GLOBAL, A new approach to computer processed log interpretation. *55th Ann. Fall Conf. SPE of AIME*, Paper SPE 9341.

MORAN, J. H. *et al.* (1962) A progress report on machine interpretation of well logs. *Trans. SPWLA 3rd Ann. Symp.*, Paper 9.

MOSELEY, L. M. (1976) Field evaluation of direct digital well logging. *Trans. SPWLA 17th Ann. Symp.*, Paper NN.

POUPON, A. and LEVEAU, J. (1976) Progress report on the automation of computer processed log data. *Trans. SPWLA 4th European Symp.*, Paper I.

RATCLIFF, J. R. *et al.* (1971) Applications of the Saraband sand-shale technique in N. America. *Trans. SPWLA 12th Ann. Symp.*, Paper P.

ROBERTS, H. and CAMPBELL, R. (1974) The application of Coriband to the evaluation of sandstone containing mica. *Trans. SPWLA 3rd European Symp.*, Paper B.

SCHMIDT, A. W. *et al.* (1972) Applications of the coriband technique to complex lithologies. *Trans. SPWLA 12th Ann. Symp.*, Paper Z.

SCOTT-SMITH, G. *et al.* (1980) CONLOG—Conco's interactive log analysis and display system. *Trans. SPWLA 21st Ann. Symp.*, Paper F.

STUDLICK, J. R. J. and GILCHRIST, W. A. J. (1981) A sensitivity study of

Schlumberger's cyberlook computation and its comparison to other evaluation methods. *Trans. SPWLA 22nd Ann. Symp.*, Paper F.

(*b*) *Interpretation*

ARCHIE, G. E. (1942) The electrical resistivity log as an aid in determining some reservoir characteristics. *Pet. Trans. AIME*, **146**, 54.

BURKE, J. A. *et al.* (1969) The lithoporosity cross plot. *Trans. SPWLA 10th Ann. Symp.*

BUSH, D. C. and JENKINS, R. E. (1977) CEC determination by correlations with absorbed water. *Trans. SPWLA 18th Ann. Symp.*, Paper H.

CLAVIER, C. and RUST, D. H. (1976) The MID plot: a new lithology technique. *Log Analyst*, **17**(6).

CLAVIER, C. *et al.* (1976) The effect of pyrite on resistivity and other logging methods. *Trans. SPWLA 17th Ann. Symp.*

CLAVIER *et al.* (1977) Theoretical and experiment basis for the dual water model, for the interpretation of shaly sands. *52nd Ann. Fall Conf. SPE of AIME*, Paper SPE 6859.

COATES, G. and DUMANOIR, J. (1974) A new approach to improved log derived permeability. *Log Analyst*, **15**(1).

EDMUNDSON, H. and RAYMER, L. L. (1979) Radioactive parameters for common minerals. *Trans. SPWLA 20th Ann. Symp.*, Vol. 1, Paper O.

FERTL, W. H. and HAMMICK, G. W. (1972) A comparative look at water saturation computations in shaly pay sands. *Log Analyst*, **13**(2).

FERTL, W. H. and TIMKO, D. J. (1971) A distinction of oil and gas in clean and shaly sands as derived from well logs. *Log Analyst*, **12**(2).

FROST, E. JR and FERTL, W. H. (1981) Integrated core and log analysis concepts in shaly clastic reservoirs. *Log Analyst*, **23**(2).

HILL, H. J. and MILLBURN, J. D. (1956) Effect of clay and water salinity on the electrochemical behaviour of reservoir rocks. *Trans. AIME*, **207**, 65.

HUSTEN, P. F. (1981) The porosity surface concept applied to water saturation equations. *Trans. SPWLA 7th European Symp.*

JUHASZ, I. (1979) The central role of Q_v and formation water salinity in the evaluation of shaly formations. *Log Analyst*, **20**(4).

KOERPERICH, E. A. (1975) Utilisation of Waxman–Smits equations for determining oil saturation on a low salinity, shaly sand reservoir. *J. Pet. Tech.*, **27** (October), 1204.

LAVERS, B. A. *et al.* (1974) Some fundamental problems of formation evaluation in the North Sea. *Trans. SPWLA 3rd European Symp.* (October).

PATCHETT, J. (1975) An investigation of shale conductivity. *Log Analyst*, **17**(6).

POUPON, A. and GAYMARD, R. (1968) Response of neutron and formation density logs in hydrocarbon bearing formations. *Log Analyst*, **9**(5).

POUPON, A. and GAYMARD, R. (1970) Evaluation of clay content from logs. *Log Analyst*, **11**(6).

POUPON, A. and LEVEAUX, J. (1971) Evaluation of S_w in shaly formations. *Log Analyst*, **12**(4).

POUPON, A. *et al.* (1970) Log analysis of sand-shale sequences—a systematic approach. *J. Pet. Tech.*, **22** (July), 867.

POUPON, A. *et al.* (1971) Log analysis in formation with complex lithologies. *J. Pet. Tech.*, **23** (August), 995.

QUIREIN, J. A. *et al.* (1981) Estimation of clay types and volume from well log data—an extension of the global method. *Trans. SPWLA 22nd Ann. Symp.*, Paper Q.

RUHAVETS, N. and FERTL, W. H. (1981) Digital shaly sand analysis based on Waxman–Smits model and log derived clay typing. *Trans. SPWLA 7th European Symp.*

SAVRE, W. C. and BURKE, J. A. (1963) Determination of more accurate porosity and mineral composition in complex lithologies with the use of sonic, neutron, and density surveys. *Trans. SPWLA 4th Ann. Symp.*

SIMANDOUX, P. (1963) Measures dielectrique en mileau poreux application a la measure des saturations en eau. Etude du Comportment des Massifs Argileux. Review de l'Institute Français du Pétrole, Supplementary Issue.

SMITS, L. J. M. (1968). SP Log interpretation in oil bearing shaly sands. *Soc. Pet. Eng. J.*, **8** (June), 123.

THOMAS, E. C. (1976) The determination of Q_v from membrane potential measurements in shaly sands. *J. Pet. Tech.*, **28** (September), 1087.

THOMAS, E. and STEIBER, S. (1975) The distribution of shale in sandstone and its effect on porosity. *Trans. SPWLA 16th Ann. Symp.*

WAXMAN, M. H. and SMITS, L. J. M. (1968) Electrical conductivities in oil bearing shaly sands. *Soc. Pet. Eng. J.*, **8** (June), 107.

WAXMAN, M. H. and THOMAS, E. C. (1974) Electrical conductivities in shaly sands. *J. Pet. Tech.*, **26** (February), 213.

WELLS, L. E. *et al.* (1979) Matrix response characteristics for sonic density and neutron. *Trans. SPWLA 20th Ann. Symp.*, Vol. 1, Paper Z.

WINSAUER, W. O. and MCCARDELL, W. M. (1953) Ionic double layer conductivity in reservoir rocks. *Trans. AIME*, **198**, 129.

WYLLIE, M. R. J. (1949) A quantitative analysis of the electrochemical component of the S.P. curve. *J. Pet. Tech.*, **1**, 17; *Trans. AIME*, **186**, 17.

(*c*) *Electromagnetic Measurements*

AJAM, S. O. (1978) The dual guard/Fo Rxo logging device. *Log Analyst*, **19**(5).

ANDERSON, W. L. (1968) Theory of borehole magnetic susceptibility measurements with coil pairs. *Geophysics*, **33**, 962–71.

BERRY, W. R. *et al.* (1979) Dielectric constant logging, a progress report. *Trans. SPWLA 20th Ann. Symp.*, Paper VV.

BROWN, R. J. S. and GAMSON, B. W. (1960) Nuclear magnetic logging. *Trans. AIME*, **219**, 199.

BROWN, R. J. S. and NEUMAN, C. H. (1980) Processing and display of nuclear magnetic logging signals: Application to residual oil determination. *Trans. SPWLA 21st Ann. Symp.*, Paper K.

CHEW, W. C. and GIANZERO, S. C. (1981) Theoretical investigation of the electromagnetic wave propagation tool. *IEE Trans. on Geoscience and Remote Sensing*, **GE19**(1).

DOLL, H. G. (1949) Introduction to induction logging and application to wells drilled with oil based mud. *J. Pet. Tech.*, **1** (June), 148; *Trans. AIME*, **186**, 148.

DOLL, H. G. (1951) A new resistivity logging method with electrodes using an automatic focussing system. *Trans. AIME*, **192**, 305.

FREEDMAN, E. and MONTAGUE, D. R. (1980) Electromagnetic propagation tool (EPT): Comparison of log derived and *in-situ* oil saturations in shaly fresh water sands. *55th Ann. Fall. Tech. Conf. SPE of AIME*, Paper SPE 9266.

FREEDMAN, R. and VAGIATZIO, J. P. (1979) Theory of microwave dielectric constant logging using electromagnetic wave propagation method. *Geophysics*, **44**(5), 969.

HERRICK, R. C. *et al.* (1979) An improved nuclear magnetism logging system and its application to formation evaluation., *54th Ann. Fall Conf. SPE of AIME*, Paper 8361.

HOYER, W. A. and RUMBLE, R. C. (1976) Dielectric constant of rocks as a petrophysical parameter. *Trans. SPWLA 17th Ann. Symp.*

KELLER, G. V. and LICASTRO, P. H. (1959) Dielectric constant and electrical resistivity of natural state cores. *USGS Bulletin*, **1052-H**, 257.

LYSNE, P. C. *et al.* (1981) Dielectric log interpretation: Probe responses and dielectric properties of rocks. *Trans. SPWLA 22nd Ann. Symp.*, Paper P.

MEADOR, R. A. and COX, P. T. (1975) Dielectric constant logging a salinity independent estimation of formation water volume. *50th Ann. Meeting SPE of AIME*, Paper 5504.

OWEN, J. E. and GREER, W. J. (1951) The guard electrode logging system. *Trans. AIME*, **192**, 347.

POLEY, J. PH. (1954) The computation of complex dielectric constant from microwave impedance measurements. *Appl. Sci. Res.*, **4B**.

POLEY, J. PH. *et al.* (1978) Use of V.H.F. dielectric measurements for borehole formation analysis. *Log Analyst*, **19**(3).

ROBINSON, J. D. *et al.* (1974) Determining residual oil with the nuclear magnetism log. *J. Pet. Tech.*, **26** (February), 226; *Trans. AIME*, **257**, 226.

SCOTT, J. H. *et al.* (1967) Dielectric constant and electrical conductivity measurement of moist rock. A new laboratory measurement. *J. Geophys. Res.*, **72**, 5101.

SCOTT, J. H. *et al.* (1981) A magnetic susceptibility well logging system for mineral exploration. *Trans. SPWLA 22nd Ann. Symp.*, Paper CC.

SEEVERS, D. O. (1966) A nuclear magnetic method for determining the permeability of sandstones. *Trans. SPWLA 7th Ann. Symp.*, Paper L.

SUAU, J. *et al.* (1972) The dual laterolog–Rxo tool. *SPE of AIME*, Paper SPE 4018.

TIMUR, A. (1968) Producible porosity and permeability of sandstones investigated through NMR principles. *Trans. SPWLA 9th Ann. Symp.*

TIMUR, A. (1969) Pulsed nuclear magnetic method for determining the permeability of sandstone. *Trans. AIME*, **246**, 775.

WHARTON, R. P. and DELANDO, J. M. JR (1981) An EPT interpretation procedure and application in fresh water, shaly oil sands. *Trans. SPWLA 22nd Ann. Symp.*, Paper E.

WHARTON, R. P. *et al.* (1980) Electro magnetic propagation logging: Advancements in technique and interpretation. *55th Ann. Fall Conf. SPE of AIME*, Paper SPE 9257.

(*d*) *Dipmeter Measurements*

CAMPBELL, R. L. (1968) Applications of dipmeter data in mid continent. *AAPG Bull.*, **52**(9), 1700.

GILREATH, J. A. and MARICELLI, J. J. (1964) Detailed stratigraphic control through dip computations. *AAPG Bull.*, **48**, 1902.
HOLT, O. R. (1972) Structural geologic considerations in diplog interpretation. *Trans. SPWLA 13th Ann. Symp.*
JOHNSON, W. M. JR and ANGEHRN, J. (1981) Advances in diplog instrumentation. *Trans. SPWLA 22nd Ann. Symp.*, Paper HH.
MORAN, J. H. *et al.* (1972) Automatic computation of dipmeter logs digitally recorded on magnetic tapes. *J. Pet. Tech.*, **24** (July), 723.
SCHOONOVER, L. G. (1974) Computer recognition of diplog patterns, a tool for stratigraphic analysis. *Trans. SPWLA 15th Ann. Symp.*, Paper X.
VINCENT, P. *et al.* (1977) Geodip: an approach to detailed dip determination using correlation by pattern recognition. *52nd Ann. Fall Conf. SPE of AIME*, Paper 6823.

(*e*) *Radioactive Logging*

ALGER, R. P. *et al.* (1972) The dual spacing neutron log (CNL). *J. Pet. Tech.*, **24** (September), 1073.
ALLEN, L. S. *et al.* (1967) Dual spaced neutron logging for porosity. *Geophysics*, **32**(1), 60.
CALDWELL, R. L. and BALDWIN, W. F. (1963) Gamma ray spectroscopy in well logging. *Geophysics*, **28**(4), 617.
CALDWELL, R. L. *et al.* (1960) Gamma radiation for inelastic scattering of 14 MeV neutron by the common earth elements. *Nuclear Science and Engineering*, **8**, 173.
CLAVIER, C. *et al.* (1971) Quantitative interpretation of thermal neutron decay time logs. *J. Pet. Tech.*, **23** (June), 743 and 756.
CULVER, R. B. *et al.* (1974) Carbon/oxygen (C/O) logging instrumentation. *Soc. Pet. Eng. J.*, **14** (October), 463.
CZUBEK, JAN A. (1971) Recent Russian and European developments in nuclear geophysics applied to mineral exploration and mining. *Log Analyst*, **12**(6).
CZUBEK, J. A. and LOSKIEWICZ, J. (1976) Optimum conditions for uranium detection in delayed neutron well logging. In *Symp. Proc. Exploration for Uranium Ore Deposits*, IAEA, Vienna, March/April.
EDMUNDSON, H. and RAYNER, L. L. (1979) Radioactive logging parameters for common minerals. *Log Analyst*, **20**(5).
FERTL, W. H. (1979) Gamma ray spectral data assists in complex formation evaluation. *Trans. SPWLA 6th European Symp.*, March.
FLAUM, C. and PIRIE, G. (1981) Determination of lithology from induced gamma ray spectroscopy. *Trans. SPWLA 22nd Ann. Symp.*, Paper H.
GARDNER, J. S. and DUMANOIR, J. L. (1980) Litho-density log interpretation. *Trans. SPWLA 21st Ann. Symp.*, Paper N.
GIVENS, W. W. *et al.* (1976) Uranium logging using a pulsed 14 MeV neutron source and detection of delayed fission neutrons. *Geophysics*, **41**(3), 468.
HEARST, J. R. and CARLSON, R. C. (1969) The RIDS: A density logger for rough holes. *Geophysics*, **34**(2), 222.
HERTZOG, R. C. (1978) Laboratory and field evaluation of an inelastic scattering and capture gamma ray spectroscopy tool. *53rd Ann. Fall Conf. SPE of AIME.*
HOPKINSON, A. H. *et al.* (1974) Depth of investigation of the neutron log. *Trans. SPWLA 15th Ann. Symp.*, Paper AA.

Hoyer, W. A. (1961) Induced nuclear reaction logging. *J. Pet. Tech.*, **13** (August), 797.

Hoyer, W. A. and Rumble, R. C. (1964) Field experience in measuring oil content, lithology and porosity with a high energy neutron induced spectral logging system. *39th Ann. Fall Conf. SPE of AIME*, Paper SPE 988.

Johnstone, C. W. *et al.* (1973) Thermal neutron decay logging using dual detection. *Log Analyst*, **14**(5).

Lawson, B. L. and Cooke, C. F. (1970) A theoretical and laboratory investigation of carbon logging Pt II—theoretical evaluation of oxygen interference. *Log Analyst*, **11**(6).

Lawson, B. L. *et al.* (1971) A theoretical and laboratory evaluation of carbon logging—laboratory evaluation. *Soc. Pet. Eng. J.*, **11**, 129.

Lock, G. A. and Hoyer, W. A. (1971) Natural gamma spectral logging. *Log Analyst*, **12**(5).

Lock, G. A. and Hoyer, W. A. (1974) Carbon–oxygen (C/O) use and interpretation. *J. Pet. Tech.*, **26** (September), 1044.

Locke, S. and Smith, R. (1975) Compensated departure curves for the thermal neutron decay time log. *Trans. SPWLA 16th Ann. Symp.*, Paper S.

Marett, G. *et al.* (1976) Shaly sand evaluation using gamma ray spectroscopy applied to the North Sea Jurassic. *Trans. SPWLA 17th Ann. Symp.*

Oliver, D. W. *et al.* (1981) Continuous carbon/oxygen (C/O) logging—instrumentation, interpretative concepts, and field applications. *Trans. SPWLA 22nd Ann. Symp.*, Paper TT.

Pickell, J. J. and Heacock, J. G. (1960) Density logging. *Geophysics*, **25**(4), 891.

Schultz, W. E. and Smith, H. D. Jr (1974) Laboratory and field evaluation of a carbon/oxygen (C/O) well logging system. *J. Pet. Tech.*, **26** (October), 1103.

Scott, J. H. (1977) Borehole compensation algorithms for a small diameter, dual detector density well logging probe. *Trans. SPWLA 18th Ann. Symp.*, Paper S.

Scott, J. H. *et al.* (1961) Quantitative interpretation of gamma ray logs. *Geophysics*, **26**(2), 182.

Serpas, C. J. *et al.* (1977) The dual detector lifetime log—theory and practical applications. *Trans. SPWLA 18th Ann. Symp.*, Paper CC.

Serra, O. *et al.* (1980) Theory, interpretation and practical applications of natural gamma ray spectroscopy. *Trans SPWLA 21st Ann. Symp.*, Paper Q.

Smith, H. D. and Schultz, W. E. (1974) Field experience in determining oil saturation from continuous C/O and Ca/Si logs, independent of salinity and shaliness. *Trans. SPWLA 15th Ann. Symp.*, Paper K.

Steinman, D. K. *et al.* (1980) 252 CF-based delayed neutron logging system for *in-situ* assay of uranium ore. *Trans. SPWLA 21st Ann. Symp.*, Paper I.

Tanner, A. B. *et al.* (1978) Uranium assay in borehole by means of germanium gamma spectrometers. *Trans. Am. Nuclear Soc.*, **28**, 112.

Tittle, C. W. (1961) Theory of neutron logging, I. *Geophysics*, **26**(1), 27.

Tittle, C. W. and Allen, L. S. (1966) Theory of neutron logging II. *Geophysics*, **31**(1), 27.

Tittman, J. and Nelligan, W. B. (1960) Laboratory studies of a pulsed neutron source technique in well logging. *Pet. Trans. AIME*, **219**, 375.

Tittman, J. and Wahl, J. S. (1965) The physical foundations of formation density logging (gamma–gamma). *Geophysics*, **30**(2), 284.

TITTMAN, J. *et al.* (1966) The sidewall epithermal neutron porosity log. *Pet. Trans. AIME*, **237**, 1351 and *J. Pet. Tech.*, **18**, 1351.

TRUMAN, R. B. *et al.* (1972) Progress report on the interpretation of the dual spacing neutron log (CNL). *Log Analyst*, **13**(4).

WAHL, J. S. *et al.* (1964) The dual spacing formation density log. *J. Pet. Tech.*, **16**, 1411.

WAHL, J. S. *et al.* (1970) The thermal neutron decay time log. *Soc. Pet. Eng. J.*, **10** (December), 365.

WICHMANN, P. A. *et al.* (1975) Field results of the natural gamma spectroscopy. *Trans. SPWLA 16th Ann. Symp.*, Paper O.

WICHMANN, P. A. *et al.* (1976) The carbon/oxygen log measurement. *Log Analyst*, **17**(6).

WILSON, W. F. and WICHMANN, P. A. (1974) The compensated neutron log and the effects of environment. *49th Ann. Fall Conf. SPE of AIME.*

WILSON, R. D. *et al.* (1980) Field evaluation of direct uranium borehole logging methods. *Trans. SPWLA 21st Ann. Symp.*, Paper S.

YOUMANS, A. H. *et al.* (1964) Neutron lifetime, a new nuclear log. *J. Pet. Tech.*, **16** (March), 319.

YOUMANS, A. H. *et al.* (1966) Neutron lifetime logging in theory and practice. *Trans. SPWLA 7th Ann. Symp.*

YOUMANS, A. H. *et al.* (1971) Applications of neutron lifetime log in new wells. *Log Analyst*, **12**(2).

(*f*) *Acoustic Logging*

ARDITTY, P. C. *et al.* (1981) First results yielded by full wave recording from a new long spacing acoustic tool. *Trans. SPWLA 7th European Symp.*, Paper 9.

ARON, J. *et al.* (1978) Formation compressional and shear interval transit time logging by means of long spacings and digital technique. *SPE of AIME*, Paper SPE 7446.

BABOUR, K. and MONS, F. (1981) Vertical seismic log: recording, processing and use. *Trans. SPWLA 7th European Symp.*, Paper 7.

BIOT, M. A. (1952) Propagation of elastic waves in a cylindrical bore containing fluid. *J. Appl. Phys.*, **23**, 997–1009.

BIOT, M. A. (1956) Theory of propagation of elastic waves in fluid saturated porous solid. *J. Ac. Soc. of America*, **28**(2).

CHRISTIANSON, D. M. (1964) A theoretical analysis of wave propagation in fluid filled drill holes for the interpretation of three dimensional velocity logs. *Trans. SPWLA 5th Ann. Symp.*, Paper K.

CHUEN HON CHENG and NAFI TOKSOZ, M. (1980) Modelling of full wave acoustic logs. *Trans. SPWLA 21st Ann. Symp.*, Paper J.

CHUEN HON CHENG *et al.* (1981) Velocity and attenuation from full waveform acoustic logs. *Trans. SPWLA 22nd Ann. Symp.*, Paper O.

COATES, G. R. and DENOO, S. A. (1980) Log derived mechanical properties and rock stress. *Trans. SPWLA 21st Ann. Symp.*, Paper U.

DOMENICO, S. N. (1976) Effect of brine/gas mixture on velocity in an unconsolidated sand reservoir. *Geophysics*, **41**(5), 882.

DUPUY *et al.* (1973) Etude numerique de l'influence de l'attenuation dans les

roches sur la forme des signaux acoustiques. *Trans. SPWLA 2nd European Symp.*
GARDNER, G. F. H. *et al.* (1964) Effects of pressure and fluid saturation on the attenuation of elastic waves in sands. *J. Pet. Tech.*, **16**(2), 189.
GEERTSMA, J. (1961) Velocity log interpretation: the effect of rock bulk compressibility. *Soc. Pet. Eng. J.*, **1** (December), 235 and 246.
GEERTSMA, J. and SMIT, D. C. (1961) Some aspects of electric wave propagation in fluid saturated porous solids. *Geophysics*, **26**(2), 169.
HARTLEY, K. B. (1981) Factors affecting sandstone compressional velocities and examination of empirical correlations between velocities and porosities. *Trans. SPWLA 22nd Ann. Symp.*, Paper PP.
HUSTEN, P. (1979) Analysis of sonic wavetrains. *Trans. SPWLA 6th European Symp.*
KOERPERICH, E. A. (1979) Shear wave velocities determined from long and short spaced acoustic devices. *SPE of AIME*, Paper SPE 8237.
KOKESH, F. *et al.* (1965) A new approach to sonic logging and other acoustic measurements. *J. Pet. Tech.*, **17** (March), 282.
LEBRETON, F. (1981) Contribution des diagraphies a la determination de la permeabilite et de la porosite dans les milieux cristallins. *Trans. SPWLA 7th European Symp.*, Paper 23.
LEBRETON, F. *et al.* (1977) Essai par diagraphie dans des milieux poreux pour evaluer l'influence de leur permeabilite sur des ondes acoustiques. *Trans. SPWLA 5th European Symp.*
PETERSON, E. W. (1974) Acoustic wave propagation along a fluid filled cylinder. *J. Appl. Physics*, **45**(8), 3340.
PICKETT, G. R. (1963) Acoustic character logs and the applications in formation evaluation. *J. Pet. Tech.*, **15**(6), 659.
RAYMER, L. L. and HUNT, E. R. (1980) An improved sonic transit time to porosity transform. *Trans. SPWLA 21st Ann. Symp.*, Paper P.
ROSENBAUM, J. H. (1974) Synthetic seismograms: logging in porous formations. *Geophysics*, **39**(1), 14.
SCOTT, J. H. and SENNA, J. (1974) Acoustic logging for mining applications. *Trans. SPWLA 15th Ann. Symp.*, Paper P.
SUMMERS, G. C. and BRODING, R. A. (1952) Continuous velocity logging. *Geophysics*, **17**(3), 598.
VOGEL, C. B. (1977) The CAD. A circumferential acoustic device for well logging. *52nd Ann. Fall Conf. SPE of AIME*, Paper SPE 6819.
WHITE, J. E. (1962) Electric waves along a cylindrical bore. *Geophysics*, **27**, 327–33.
WHITE, J. E. and ZECHMAN, R. E. (1968) Compensated response of an acoustic logging tool. *Geophysics*, **33**(2), 302.
WILEY, R. (1980) The borehole televiewer—Revisited. *Trans. SPWLA 21st Ann. Symp.*, Paper HH.
WYLLIE, M. R. J. *et al.* (1956) Elastic wave velocities in heterogeneous and porous media. *Geophysics*, **21**(1), 41.
WYLLIE, M. R. J. *et al.* (1958) An experimental investigation of factors affecting wave velocities in porous media. *Geophysics*, **23**(3), 459.
WYLLIE, M. R. J. *et al.* (1962) Studies of electric wave attenuation in porous media. *Geophysics*, **27**(5), 569.

(*g*) *Cement Bond Log*

Brown, H. D. *et al.* (1971) New developments in sonic wave train display and analysis in cased holes. *Log Analyst*, **12**(1).

Grosmangin, M. *et al.* (1961) A sonic method for analysing the quality of cementation of borehole casings. *J. Pet. Tech.*, **13** (February), 165.

McGhee, B. F. and Vacca, H. L. (1980) Guidelines for improving monitoring of cementing operations. *Trans. SPWLA 21st Ann. Symp.*, Paper V.

Pardue, G. H. *et al.* (1963) Cement bond log—a study of cement and casing variable. *J. Pet. Tech.*, **15**(5), 545.

Pickett, G. R. (1966) Prediction of interzone fluid communication behind casing by use of the cement bond log. *Trans. SPWLA 7th Ann. Symp.*

Pilkington, P. E. and Fertl, W. H. (1975) Field tests of cement bond logging tools. *Log Analyst*, **16**(4).

Walker, T. (1968) A full wave display of acoustic signal in cased holes. *J. Pet. Tech.*, **20** (August), 811.

(*h*) *Fracture Location and Characterisation*

Aguilera, R. N. (1976) Analysis of naturally fractured reservoirs from conventional well logs. *J. Pet. Tech.*, **28**, 1184.

Aguilera, R. N. and van Poollen, H. K. (1977) Current status on the study of naturally fractured reservoirs. *Log Analyst*, **18**(3).

Albright, J. N. *et al.* (1980) Transmission of acoustic signals through fractures. *Trans. SPWLA 21st Ann. Symp.*, Paper R.

Beck, J. *et al.* (1977) Reservoir evaluation of fractured cretaceous carbonate in South Texas. *Trans. SPWLA 18th Ann. Symp.*, Paper M.

Brown, R. O. (1978) Application of fracture identification logs in the cretaceous of North Louisiana and Mississippi. *Trans. Gulf Coast Assoc. of Geol. Soc.*, **28**.

Delhomme, J. P. *et al.* (1981) An integrated approach to fracture detection from multipad sensors. *Trans. SPWLA 7th European Symp.*, Paper 26.

Granger, P. and McCann, D. (1977) Interborehole acoustic measurements in site investigation. *Quart. J. Eng. Geol.*, **10**, 241.

Heflin, J. D. (1979) Fracture detection in west coast reservoirs using well logs. *SPE of AIME*, Paper SPE 7976.

Heflin, J. D. and Frost, E., Jr (1981) Some novel approaches in the use of wireline logs for fracture detection. *Trans. SPWLA 22nd Ann. Symp.*, Paper ZZ.

Keys, W. S. (1979) Borehole geophysics in igneous and metamorphic rocks. *Trans. SPWLA 20th Ann. Symp.*

Koerperich, E. A. (1978) Investigation of boundary waves and interference patterns as techniques for detecting fractures. *J. Pet. Tech.*, **30**, 1199.

Paillet, F. L. (1980) Acoustic propagation in the vicinity of fractures which intersect a fluid filled borehole. *Trans. SPWLA 21st Ann. Symp.*, Paper DD.

Pickett, G. R. (1963) Acoustic character logs and their application to formation evaluation. *J. Pet. Tech.*, **15**, 659.

Pickett, G. R. and Reynolds, E. B. (1969) Evaluation of fractured reservoirs, *Soc. Pet. Eng. J.*, **9**, 28.

Suau, J. and Gartner, J. (1980) Fracture detection from well logs. *Log Analyst*, **11**(2).

TATE, T. K. *et al.* (1970) The hydrogeological investigation of fissure flow by borehole logging techniques. *Quart. J. Eng. Geol.* (February), 195.
VOGEL, C. B. (1977) The CAD, circumferential acoustic device for well logging. *SPE of AIME*, Paper 6819.
ZEMANEK, J. *et al.* (1969) The borehole televiewer—a new concept for fracture location and other types of borehole inspection. *J. Pet. Tech.*, **21**, 762.

(*i*) *Measurements While Drilling*

ARPS, J. J. and ARPS, J. L. (1964) The sub-surface telemetry problem—a practical solution. *J. Pet. Tech.*, **16** (May), 487.
BURGESS, T. M. and LESSO, W. G. (1985) Measuring the wear of milled tooth bits using MWD torque and weight on bit. SPE Paper SPEIIADC 13475.
CAHILL, J. (1966) Use of the drill pipe electric log. *Trans. SPWLA 7th Ann. Symp.*
HELANDER, D. P. (1966) Here are ways you can log formations while drilling. *Oil and Gas Equipment* (September).
MCDONALD, W. J. and WARD, C. E. (1975) Borehole telemetry system is the key to continuous downhole drilling measurements. *Oil and Gas Journal*, **73**(37), 111.
MCDONALD, W. J. and WARD, C. E. (1976) Logging while drilling—a survey of methods and priorities. *Trans. SPWLA 17th Ann. Symp.*, Paper U.

(*j*) *Geopressure and Mechanical Properties*

ANDERSON, R. A. *et al.* (1972) Fracture pressure gradient determination from well logs. *47th Ann. Fall Conf. SPE of AIME*, Paper SPE 4135.
COATES, G. R. and DENOO, S. A. (1980) Log derived mechanical properties and rock stress. *Trans. SPWLA 21st Ann. Symp.*, Paper U.
COATES, G. R. and DENOO, S. A. (1981) Mechanical properties program using borehole analysis and Mohr's circle. *Trans. SPWLA 22nd Ann. Symp.*, Paper DD.
FERTL, W. H. (1975) *Abnormal Formation Pressures: Developments in Petroleum Science*, Elsevier Scientific Publishing Co., Amsterdam.
HOTTMAN, C. E. and JOHNSON, R. K. (1965) Estimation of formation pressure from log derived shale properties. *J. Pet. Tech.*, **17** (June), 717.
MATTHEWS, W. R. and KELLY, J. (1967) How to predict formation pressure and fracture gradient. *Oil and Gas Journal*, February 20th, p. 92.
REYNOLDS, E. B. *et al.* (1972) Potential hazards of acoustic log–shale pressure plots. *47th Ann. Fall Conf. SPE of AIME*, Paper SPE 4020.
STEIN, N. (1975) Mechanical properties of friable sands from conventional log data. *J. Pet. Tech.*, pp. 757–63; *50th Ann. Fall Conf. SPE of AIME*, Paper SPE 5500.
TIMKO, D. J. (1965) Practical applications of log calculated pressures as an aid to solving drilling problems. *Log Analyst*, **6** (October).
TIXIER, M. P. *et al.* (1973) Estimation of formation strength from the mechanical properties log. *48th Ann. Fall Conf. SPE of AIME*, Paper SPE 4532.
WALLACE, W. F. JR (1965) Predicting formation pressures from well logs. *Log Analyst*, **6** (March).
WICHMANN, P. A. (1974) A review of the use of logs to determine abnormal pressures. *Trans. SPWLA 15th Ann. Symp.*, Paper C.

(*k*) *Borehole Gravity Measurements*

CATON, P. W. (1981) Improved methods for reducing borehole gravity data—applications and analysis of reduced gravity plots. *Trans. SPWLA 22nd Ann. Symp.*, Paper RR.

GOODELL, R. R. and FAY, C. H. (1974) Borehole gravity meter and its application. *Geophysics*, **29**, 774–82.

HEARST, J. R. (1977) On the range of investigation of a borehole gravimeter. *Trans. SPWLA 18th Ann. Symp.*, Paper E.

HOWELL, L. G. *et al.* (1966) The development and use of a downhole high precision downhole gravity meter. *Geophysics*, **31**, 764–72.

JAGELER, A. M. (1976) Improved hydrocarbon reservoirs evaluation through use of borehole-gravimeter data. *J. Pet. Tech.*, **28**, 709–18.

JONES, W. R. (1972) The use of downhole gravity data in formation evaluation. *Log Analyst*, **13**(6) and *Trans. SPWLA 13th Ann. Symp.*

LAURIN, P. J. *et al.* (1981) Borehole gravimetry: an application example to civil engineering. *Trans. SPWLA 7th European Symp.*, Paper 27.

McCULLOGH, T. H. (1965) A confirmation by gravity measurements of an underground density profile based on core densities. *Geophysics*, **30**, 1108–32.

McCULLOGH, T. H. *et al.* (1968) Application of gravity measurement in wells to problems in reservoir evaluation. *Trans. SPWLA 9th Ann. Symp.*

RASMUSSON, N. F. (1973) Borehole gravity survey planning and operations. *Trans. SPWLA 14th Ann. Symp.*

RASMUSSON, N. F. (1975) The successful use of the borehole gravity meter in Northern Michigan. *Log Analyst*, **16**(5).

RASMUSSON, N. F. (1977) Gravity logs promise new applications. *Pet. Engineer* (July), 21–4.

ROBBINS, S. L. (1981) Re-examination of the values used as constants in calculation of rock density from borehole gravity data. *Geophysics*, **46**, 208–10.

SCHMOKER, J. W. (1978) Accuracy of borehole gravity data. *Geophysics*, **43**, 538–42.

SNYDER, D. D. (1976) The borehole gravity anomaly—application in interpreting borehole gravity surveys. *Trans. SPWLA 17th Ann. Symp.*

(*l*) *Formation Testing and Production Monitoring*

CMELIK, H. and SARABIAN, A. R. (1979) Quantitative analysis of production logs in two phase liquid–gas systems. *SPE Production Technology Symp.* (Lubboek), Paper SPE 8761.

LEACH, B. C. *et al.* (1974) The full bore flowmeter. *49th Ann. Fall Conf. SPE of AIME*, Paper 5089.

MILBURN, J. D. and HOWELL, J. C. (1961) Formation evaluation with the wireline tester—merits and shortcomings. *J. Pet. Tech.*, **13** (October), 987.

MORAN, J. H. and FINKLEA, E. E. (1962) Theoretical analysis of pressure phenomena associated with the wireline formation tester. *J. Pet. Tech.*, **14** (August), 899.

PETOVELLO, B. G. (1975) Evaluation of well performance through production logging. *Trans. SPWLA 16th Ann. Symp.*, Vol. 1, Paper E.

SARABIAN, A. R. (1981*a*) Pressure gradient determination from the fluid density log. *SPE Deep Drilling and Production Symp.* (Amarillo), Paper SPE 9700.

SARABIAN, A. R. (1981*b*) Gas condensate evaluation in multi-layered systems by production logging tools. *Trans. SPWLA 7th European Symp.*, Paper 32.

SETHI, D. K. *et al.* (1980) The formation multi tester, its basic principles and practical field application. *Trans. SPWLA 21st Ann. Symp.*, Paper CC.

VAN ROOY, D. L. and VESPERMAN, F. A. (1981) A generalised, constrained N-pass method for interpreting FBS and CFS flowmeters. *Trans. SPWLA 22nd Ann. Symp.*, Paper W.

WOOD, B. and CAMPBELL, R. (1974) Gas well completion evaluation with the production combination tool. *Trans. SPWLA 3rd European Symp.*, Paper K.

(*m*) *Non-Petroleum Logging*

(*1*) *Coal*

BOND, L. O. *et al.* (1971) Well log applications in coal mining and rock mechanics. *Trans. AIME*, **250** (December), 355.

BROM, R. W. C. and DRIEDONKS, F. (1981) Applications of petrophysical logging in the evaluation of coal deposits. *Trans. SPWLA 22nd Ann. Symp.*, Paper KK.

LAVERS, B. A. and SMITS, L. J. M. (1977) Recent developments in coal petrophysics. *Log Analyst*, **13**(1).

PETERS, M. and KEMPTON, N. H. (1977) Wireline logging for coal exploration in Australia. *Log Analyst*, **13**(3).

(*2*) *Engineering geology*

BUCHAN, S. *et al.* (1972) Relations between the acoustic and geothermal properties of marine sediment. *Quart. J. Eng. Geol.*, **5**, 265–84.

MCCANN, D. M. and MCCANN, C. (1977) The application of borehole acoustic logging techniques in engineering geology. *Log Analyst*, **18**(3).

MYUNG, J. I. and HELANDER, D. P. (1972) Correlation of elastic moduli dynamically measured by *in-situ* and laboratory techniques. *Trans. SPWLA 13th Ann. Symp.*

(*3*) *Geothermal*

BENOIT, W. R. *et al.* (1980) Geothermal well log analysis at Desert Peak, Nevada. *Trans. SPWLA 21st Ann. Symp.*, Paper AA.

DENNIS, H. R. (1980) Borehole survey instrumentation development for geothermal appliances. *Trans. SPWLA 21st Ann. Symp.*, Paper D.

KELLER, G. V. *et al.* (1975) Geophysical logs from the Kilauea geothermal research drill hole. *Trans. SPWLA 15th Ann. Symp.*, Paper L.

MATTHEWS, M. (1980) Calibration models for fractured igneous rock environments. *Trans. SPWLA 21st Ann. Symp.*, Paper L.

VENERUSS, A. F. and COQUAT, J. A. (1979) Technology development for high temperature logging tools. *Trans. SPWLA 20th Ann. Symp.*, Paper KK.

Chapter 4

PRESSURE ANALYSIS: THE HARDWARE

E. DIEMER

Flopetrol Johnston Schlumberger, Melun, France

1. INTRODUCTION

Being highly sensitive to all events, pressure—its value and its variation—is a parameter of prime importance for knowledge of an oil reservoir. In Chapter 5 all the information which can be derived from the correlation of pressure with flow rate will be described. In this chapter, the technology used in the oil field and the performance which can be expected from gauges currently available on the market will be described.

2. PRESSURE GAUGE TECHNOLOGY

2.1. Pressure Gauges and Transducers

Every pressure gauge consists of three parts:

—the transducer
—the transmitter
—the recorder

The transducer is the basic element: exposed to pressure stresses, the geometrical or intrinsic characteristics of the transducer are modified. The function of the transmitter is to transform the pressure-induced variations into measurable data. These data can be recorded downhole by various means or transmitted through an electric line to a recorder situated at the surface.

2.2. Description of Typical Transducers

A pressure-sensing device subjected to a pressure stress undergoes geometrical or mechanical variations. These variations can be monitored either by a mechanical system (mechanical gauges) or by generating an electrical signal (electronic gauges).

An exhaustive description of all the principles which have been used in the design of pressure transducers will not be given; only those which have been or are currently used in petroleum exploration and production will be considered.

2.2.1. The Bourdon Tube

The most widely used pressure gauges are based on a Bourdon tube: a hollow cylindrical tube is closed at one end and exposed to pressure through the other end. The general shape of the tube is more or less like a C; when submitted to pressure, the C expands (Fig. 1) and any dial attached to the closed end of the tube can monitor the pressure variation. In the oil-field industry, helical Bourdon tubes are currently used (see Section 3.1.1.).

2.2.2. Strain Gauge Transducers

The transduction element in the strain gauge transducer is a resistor which is mounted to the pressure-sensing device so as to cause the resistor to be sensitive to strain. When the pressure-sensing device undergoes a displacement, the strain-sensitive resistor changes its physical length thereby causing a change in resistance. This effect may be expressed as

$$GF = \frac{\Delta R/R}{\Delta L/L}$$

where GF = gauge factor, ΔR = the change in resistance, R = the unstrained resistance, ΔL = the change in length and L = the unstrained length. Various types of strain gauge transducers have different gauge factors.

We will describe three types of strain gauge transducers: their corresponding gauge factors are given below. A rule that applies to these transducers is: the larger the gauge factor, the higher the output of the device.

Type	*Gauge factor*
Unbonded wire	4
Thin film	2
Semiconductor	80–150

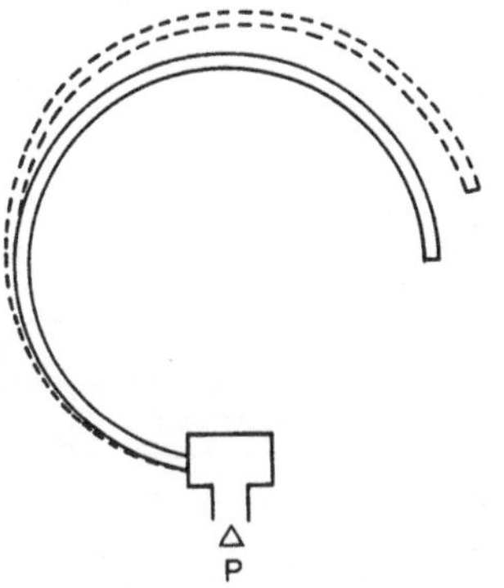

FIG. 1. Bourdon tube.

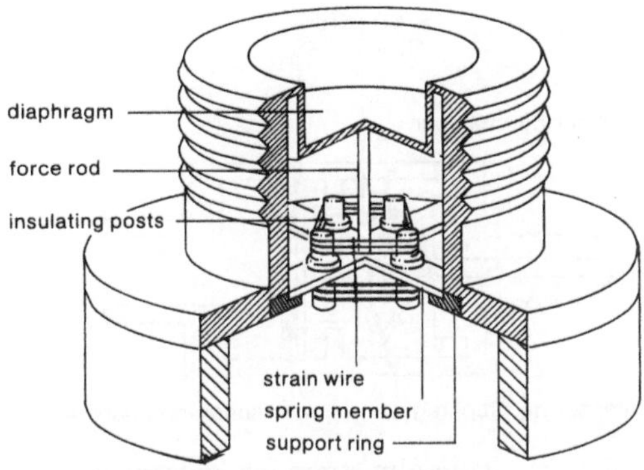

FIG. 2. Cut-away of an unbonded strain gauge. (Courtesy of Bell & Howell.)

2.2.2.1. Unbonded wire transducer. The unbonded wire strain gauge transducer, Fig. 2, consists of wire wrapped around a mechanical assembly which causes the wire to stretch or contract in proportion to the displacement of the force-summing device.

2.2.2.2. Thin film strain gauge (Fig. 3). One side of the diaphragm is subjected to pressure which can be measured by resistors attached to the other side. The resistors are deposited by a thin film technique (evaporation under vacuum or sputtering procedures).

Any pressure variation ΔP leads to a microdeformation of the diaphragm Δx, this deformation being proportional to ΔP. The resistors, which are intimately bonded to the diaphragm, are subjected to some deformation which leads to a variation ΔR of their initial value. Thus,

$$\Delta R/R \cong \Delta P$$

These resistors are interconnected in a Wheatstone bridge. The resulting signal is a voltage, which is proportional to the pressure:

$$V = E\frac{\Delta R}{R} \cong \Delta P$$

As the behaviour of the diaphragm/resistor assembly is sensitive to temperature, any temperature variation must be taken into account; for accurate measurements, the temperature is separately measured, but many manufacturers incorporate compensation components in their transducers, the essential aim of which is to compensate to a certain extent for temperature variations.

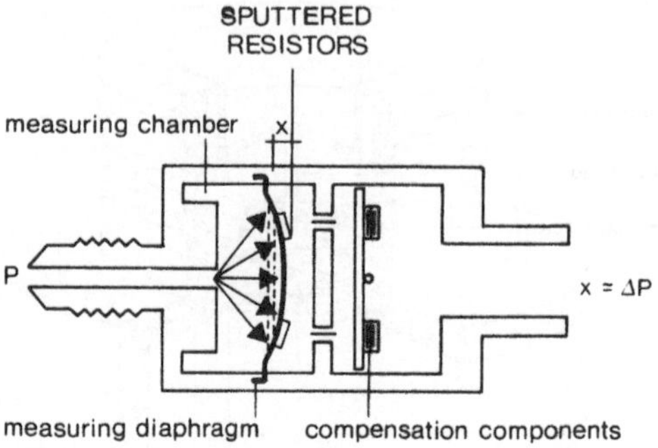

FIG. 3. Thin film strain gauge transducer.

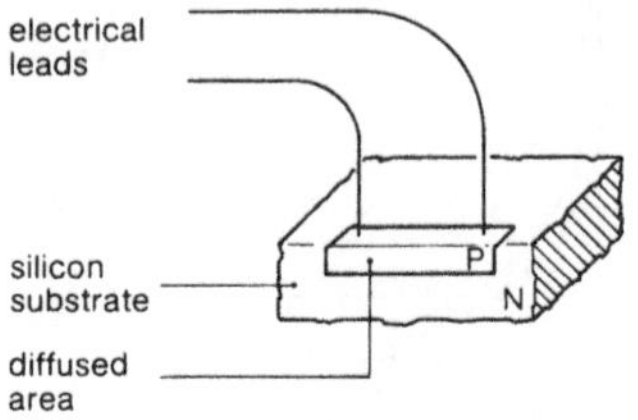

FIG. 4. Semiconductor transducer.

2.2.2.3. Semiconductor transducer. The semiconductor strain gauge transducer, Fig. 4, consists of a piezoresistive element diffused directly into bulk silicon material. This element is then mechanically coupled to a force-summing element such as a diaphragm. The principal advantages of this gauge are low cost (due to automation of the transducer manufacturing process) and high output level. These transducers can be constructed with integral amplifiers which give them a high output level (10 V) and a low output impedance. The main disadvantage of this type of gauge is the temperature limitation for oil industry application (110–125 °C). Semiconductor technology could become a significant factor in future transducer performance.

2.2.3. Capacitive Transducer

Figure 5 shows a capacitive-type transducer. Pressure displacing the force-summing element in this transducer causes a change in capacitance. If a diaphragm is used, it may be one plate of the capacitor, or if a Bourdon tube is used, a moveable plate may be mechanically coupled to the end.

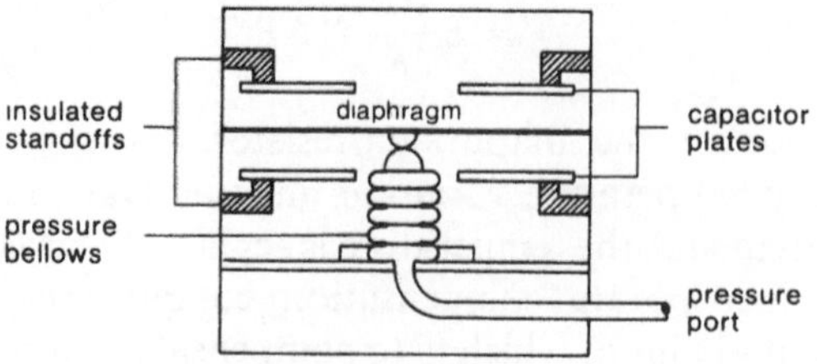

FIG. 5. Capacitive transducer.

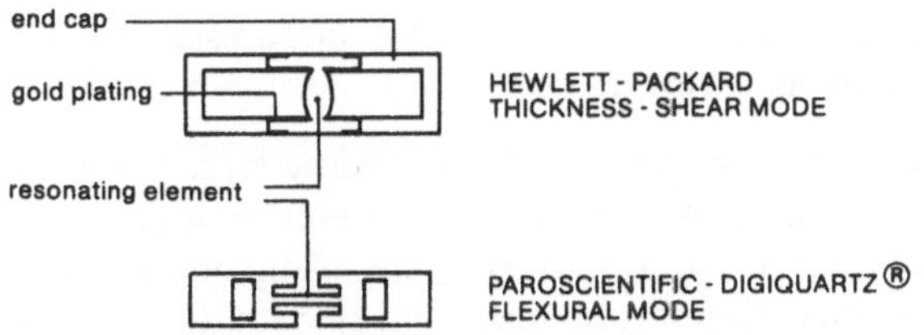

FIG. 6. Vibrating crystal transducer.

Regardless of the method used, the force measured is reflected by a change in capacitance. The basic output of the capacitive transducer is a frequency signal, which digital circuits can use directly.

2.2.4. Vibrating Crystal Transducer

In vibrating crystal transducers (Fig. 6), a quartz crystal is forced by external electronic circuits to oscillate at its resonating frequency. When a stress is applied to the resonator, its resonating frequency shifts proportionally to the stress. In the Hewlett Packard crystal system, the pressure-sensing element is the crystal itself, whereas in the Digiquartz® crystal produced by Paroscientific, a lever/rod system transmits the pressure stress to the resonating part.

2.3. Downhole versus Surface Recording

Historically, the first downhole gauges used were self-contained, and therefore they recorded downhole; this was in fact the only possibility available with mechanical gauges some of which will be described in Section 3.1.

Electronic gauges generate an electrical signal and, here, the pressure data can either be recorded downhole or transmitted to the surface through an electric cable. Surface recording brings an enormous advantage with the possibility of monitoring in real time, at surface, the behaviour of the well and then of making any decision which may be necessary as a result. Most gauges transmit the information to the surface in digital data, the quality of which is therefore independent of cable characteristics, such as insulation. Pressure data are scanned at preselected intervals, for instance 1 s, stored on a cassette or a floppy disk, and plotted and/or printed in adequate engineering units on a computerised data acquisition system.

But in spite of the obvious advantages of having data available in real time at surface, there are still a few circumstances where downhole recording is more advisable:

—The section of the electrical cable being larger than the current slick line [$\simeq$5·6 mm (0·219 in) in diameter compared with $\simeq$2·1 mm (0·082 in) for a typical slick line], the lift up on the cable is more than 7 times higher; this means more sinker bars, longer lubricator, and in addition a heavier and larger winch, and a heavier and more sophisticated stuffing box system. In other words, cost, space availability or transportation problems may hinder the use of an electrical cable.

—When corrosion is a problem, technicians may not like to let a cable be exposed to sour gas for many hours. In this case, a downhole recording gauge can be set in the bottom of the well, the cable removed and the gauge retrieved once the test has been completed.

—For safety reasons, some companies do not like to have a cable in a well during a well test operation when working from a floater.

For all these reasons, downhole recording is still preferred by many engineers. But, even in this case, electronic recording gauges are much more flexible than mechanical recording gauges. However, their construction only became possible at the time when adequate storage capacity appeared on the market (see Section 3.1.3).

3. DESCRIPTION OF TYPICAL GAUGES

3.1. Downhole Recording Gauges

Self-contained pressure gauges, run on slick-line have been used for many decades in the petroleum industry. These gauges consist of three basic components:

—a transducer or pressure-sensing device
—a recorder
—a clock

Such systems enable the recording of pressure versus time.

3.1.1. Mechanical Gauges

As far as mechanical gauges are concerned, three makes have been used and are still in use in the oil-field industry: Amerada, Kuster and Leutert gauges.

3.1.1.1. Amerada/Kuster gauges (Fig. 7). The $1\frac{1}{4}$ in OD Amerada RPG-3 gauge is by far the most commonly used in the oil industry (the first one, built in 1929, was called the Amerada bomb).

The pressure-active element is a helical Bourdon tube fixed at one end and free to rotate at the other end, the inside of which is subjected to the well pressure. The resulting rotation of the free end of the Bourdon tube is transmitted directly to a recording stylus without the use of gears or levers. The stylus records onto a black-coated metal chart, which is carried by a removable chart holder connected to a clock. As time passes, the chart is progressively moved down and the stylus scratches a very fine thin line showing pressure variation versus time. Once removed from its holder, the chart appears as a 5 in × 2 in diagram whose length is the time scale, and whose width is the pressure scale. The 2 in corresponds to a 320° rotation of the stylus, which is the maximum twist of the Bourdon tube.

The extreme ranges covered by standard Bourdon tubes are 0–500 psi and 0–25 000 psi; many other intermediates exist such as 0–1000 psi, 0–2000 psi, 0–5000 psi, etc. Clocks can be used in the range from 3 to 360 h, but they are seldom used above 120 h. Each clock can also be fitted with a double pitch which enables a 3-h clock, for example, to be transformed into a $1\frac{1}{2}$-h clock. A current practice, when the build-up is long but early-time analysis is recommended, is to run two gauges simultaneously, one equipped with a short-range clock, the other with a long-range clock.

In order to protect the inside of the Bourdon tube from the well fluid, which could be corrosive, it is filled with silicon oil and sealed with bellows which, in fact, transmit pressure. Although compensated for temperature effects, accurate measurements require calibration to be made at various temperatures and the reading of stylus deflection has to be converted into pressure using the coefficient for the appropriate temperature. The temperature can be measured using a maximum reading thermometer housed inside the thermometer well.

It is also worth noting that the same Bourdon tube assembly can be used to record the temperature: in this case, the bellows are replaced by a solid reservoir containing a volatile liquid. The vapour pressure of the volatile liquid is directly related to its temperature, and using appropriate calibration factors enables the temperatures to be calculated (Fig. 7).

The stated accuracy of the Amerada pressure gauge is ±0·25% of full scale (±25 psi) and the sensitivity for a well fixed gauge 0·05% (±5 psi for a 10 kpsi). Although higher resolution is impossible to get, a better accuracy can be obtained by calibrating the chart in the Amerada gauge and then running the gauge into the well with this specific chart still in place. Thus,

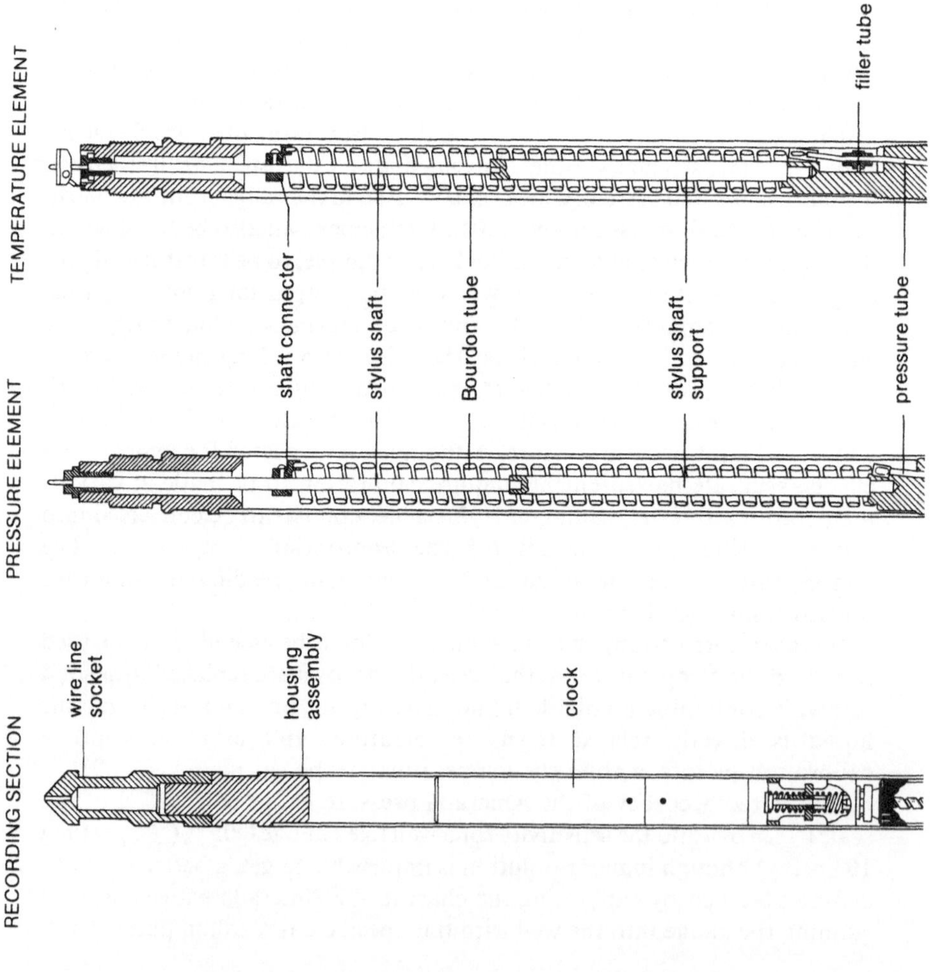
RECORDING SECTION
wire line socket
housing assembly
clock
PRESSURE ELEMENT
TEMPERATURE ELEMENT
shaft connector
stylus shaft
Bourdon tube
stylus shaft support
pressure tube
filler tube

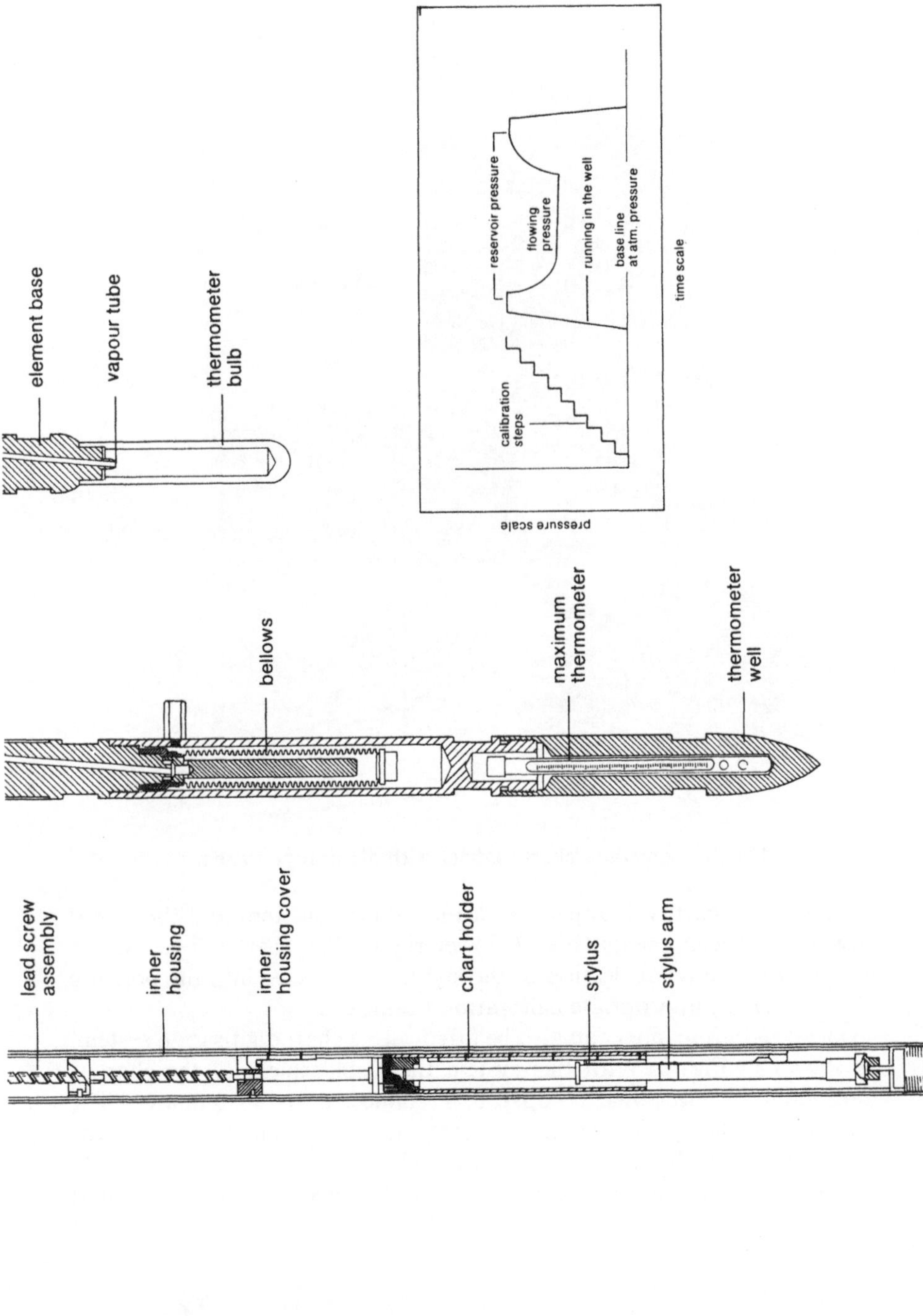

FIG. 7. Amerada RPG-3 gauge and typical chart. (Courtesy of Geophysical Research Corp.)

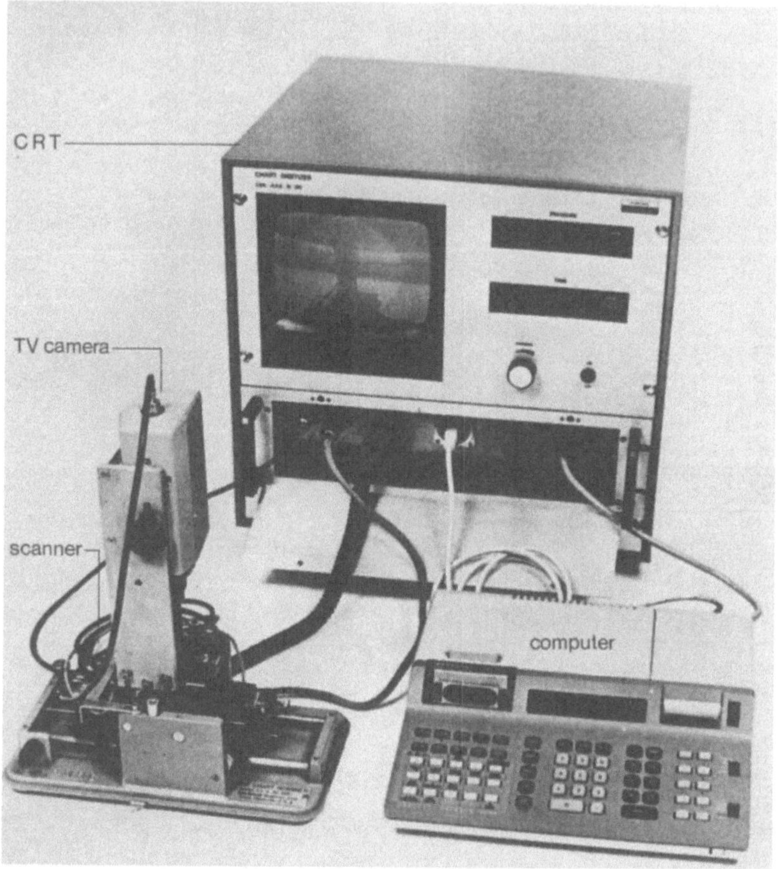

FIG. 8. Amerada chart scanner with digitisation system.

on the same chart, will appear both the calibration steps and the actual downhole measurements (Fig. 7, lower right). The chart is read with a microscope and the deflection of the stylus is converted into the pressure value using the appropriate calibration coefficient.

The standard scanner can also be fitted with a chart digitisation system, as shown in Fig. 8. With this system the displacement of the chart is monitored by electronic comparators connected to a computer. The operator starts by loading the program into the computer giving the specification of the gauge to be read: clock used, drift of the clock, temperature, calibration coefficient, etc. After installation of the chart under the microscope, the operator assures the coincidence between the cross-wire and the pressure line; whenever he wants a pressure point to be calculated, he pushes the 'execute' button of the computer which will calculate time and pressure for this given point. Chart reading can be made either by looking through the microscope or by using a camera and a TV monitor, as shown in Fig. 8.

Amerada gauges are produced by the Geophysical Research Corporation, but very similar instruments are built by Kuster Company, many parts being exchangeable between the two makes. Figure 9 presents a cut-away view of such a gauge. Although the most used instrument has a $1\frac{1}{4}$ in OD, some special applications require different size instruments as shown in Fig. 10.

3.1.1.2. Leutert gauge (Fig. 11). In the Leutert precision subsurface pressure gauge, a piston is subjected to the wellbore pressure which causes it to move against the tension of a calibrated helical spring. The stylus attached to a rod, part of the piston, is protruded along the long side of the diagram while the diagram is rotated versus time (3–12 h per full rotation). The diagram is of larger size than that of the Amerada; the length is the pressure scale and the width is the time scale.

The accuracy given by the Leutert gauge is stated as being greater than the Amerada gauge, but since the diagram makes many turns during a typical operation, the overlappings of the line drawn by the stylus are sometimes difficult to interpret.

3.1.2. Electromechanical Gauges

Since mechanical friction may reduce the sensitivity of the instrument, a few manufacturers have produced a frictionless transmission of the stylus. This is particularly the case for Sperry Sun's Bourdon tube pressure gauge. The deformation of the Bourdon tube is transmitted to the stylus without any mechanical connection, using a battery-powered electronic programmer. Here, the accuracy stated is $\pm 0{\cdot}05\%$ (± 5 psi for 10 000 psi), and the resolution is $0{\cdot}005\%$ ($\pm 0{\cdot}5$ psi).

3.1.3. Electronic Gauges

The pressure-sensing device, although basically still mechanical—since it undergoes geometrical or mechanical variations generated by pressure—is coupled to a transduction device which translates these variations into an electrical signal. The transducer modules used in the downhole recording gauges are not specifically designed for the downhole recording function

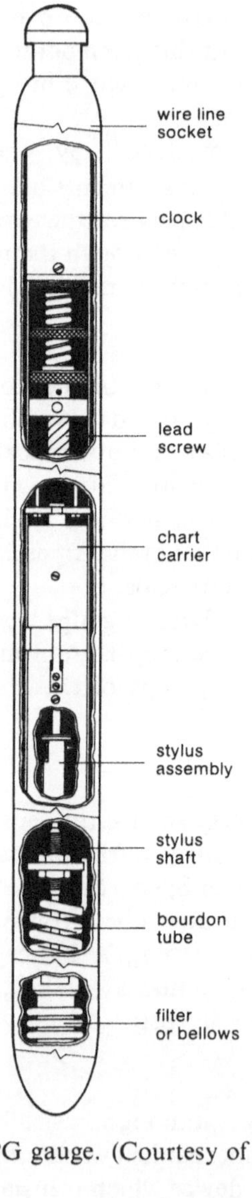

FIG. 9. Kuster KPG gauge. (Courtesy of Kuster Company.)

GEOPHYSICAL RESEARCH CORPORATION	KUSTER C°	OD"	APPLICATION
RPG 3 L = 74 or 78"	KPG L = 73⅜"	1¼"	standard use (best accuracy)
	K-3 L = 47. 4" or 41.3"	1¼"	production or drill stem testing
RPG 4 L = 71⅞" or 75⅜"	K-2 L = 39.1" or 44.6"	1"	in small tubing and production or drill stem testing
	K-4 L = 42"	3/4"	in macaroni tubing
RPG 5 L = 18"	K-5 L = 19.25"	1.5"	to be used in pump down or through flowline technique
	AK-1 L = 39"	2¼"	rugged instrument for testing string

FIG. 10. Various GRC and Kuster gauges.

but can also be used with an electric line. In addition, some recording systems can accommodate various types of sensor modules. For this reason, essentially, the downhole processor will be described here and the sensor modules will be discussed in Section 3.2.

Three downhole recording gauges are currently on the market. We will describe one of them and just mention the two others.

The Solid State Downhole Processor (SSDP(TM)) uses either a strain gauge or a quartz crystal gauge as sensing element. The data are recorded on static memories. The first trials were made in 1973 by Flopetrol Johnston (a Schlumberger Company) to store data on non-volatile memory chips. Close cooperation with the manufacturer enabled their application to be extended and, in particular, a high temperature writing-in procedure to be settled. Based on these memories, the first tool which appeared in the field in 1978 was the SSDR(TM): Solid State Downhole Recorder. It made use of EPROMs (UV Erasable Programmable Read Only Memories).

But development in advanced technology made the replacement of this tool rapidly possible by a more powerful version: the SSDP, which makes use of microprocessors and has built-in processing capabilities. The desired program is loaded prior to running-in. A data reduction algorithm enables only significant data to be kept in memory. Thanks to this algorithm, a scanning rate of 1 second can be used without incurring the risk of not being able to store all the data because of the memory chips being filled up. Simulations have shown that, thanks to this data compression algorithm which, by construction, can attain 1/512, 500-h tests can be recorded with a 1-second scanning rate. The pressure data can

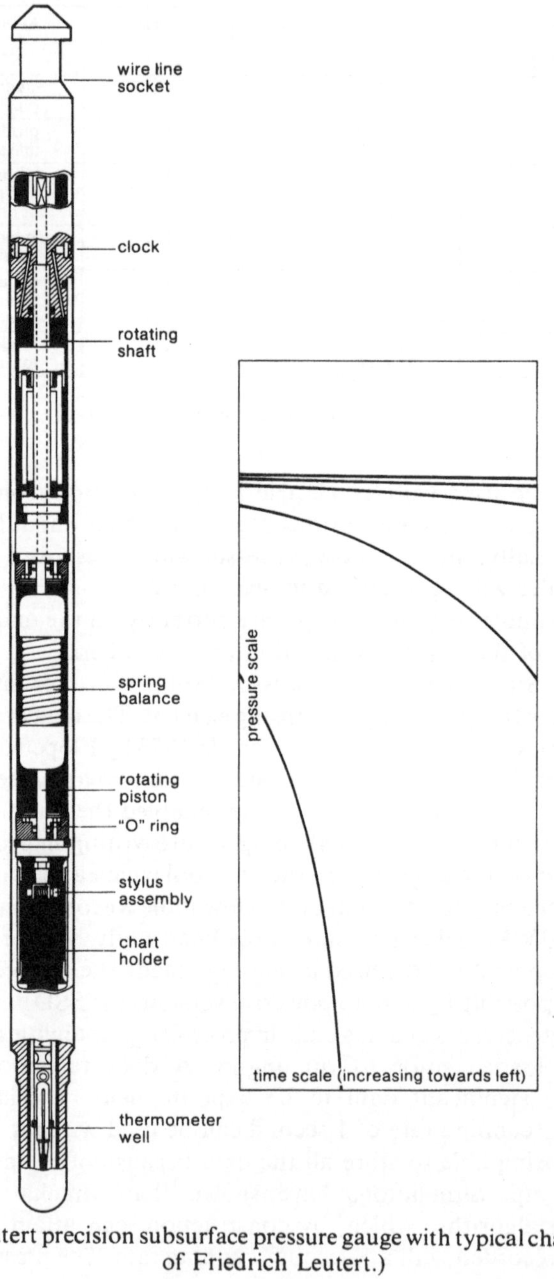

FIG. 11. Leutert precision subsurface pressure gauge with typical chart. (Courtesy of Friedrich Leutert.)

be either scanned instantaneously, or averaged over a chosen number of measurements.

Delay time enabling the recording to start only after a programmed duration, is also incorporated. This would avoid, for example, recording insignificant data during the running-in of a DST. This tool can also be used for recording during very long periods (3 to 4 weeks). This is achieved by the tool being put on steady state—which saves battery life—and reactivated according to a programmed scanning rate.

Figure 12 shows the block diagram of this SSDP. You will note on this schematic, the existence of a ROM (Read Only Memory) containing the calibration coefficients; so, when reading the pressure, the first step the program does is to load these coefficients into the processing computer,

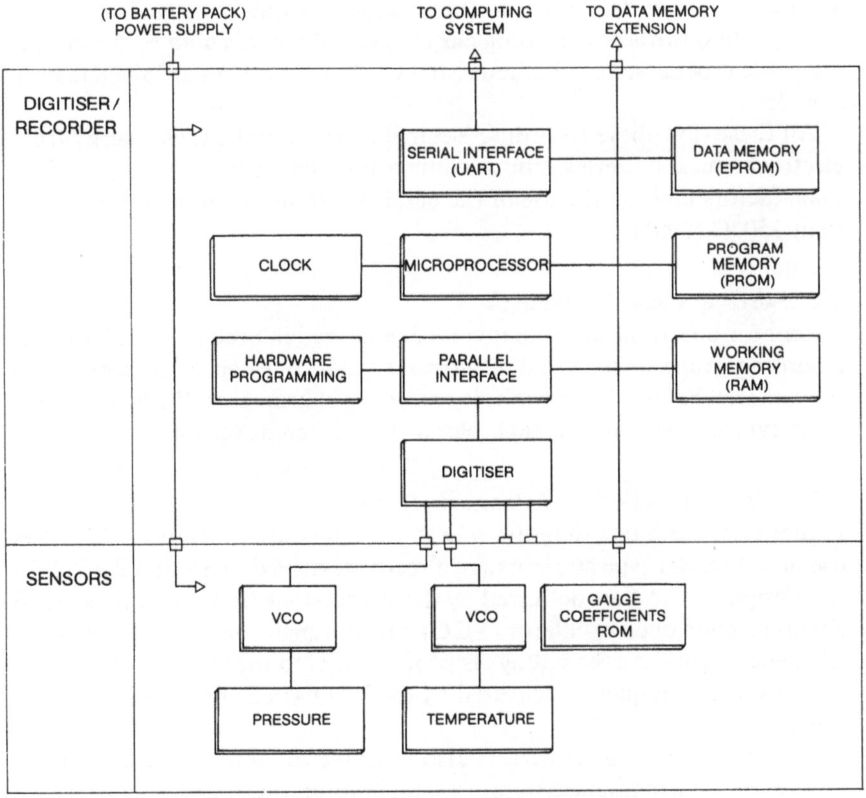

FIG. 12. SSDP tool—general block diagram.

thus avoiding human error in transferring the coefficients into the program (about 8 to 15 coefficients are normally used, with 7 digits each).

New progress in memory technology has led to the development of EEPROMs (Electrically Erasable Programmable Read Only Memories). Although their number of bits/chip ratio is still lower than for the EPROMs, they have two advantages over the latter:

1. they are easier to erase once the job is completed (electrical pulse only);
2. they are more appropriate for writing-in and reading at high temperature.

These new memories will, that way, progressively replace the EPROMs.

Lynes Company is producing a tool which uses this type of memory and can accommodate either a strain gauge or a quartz gauge transducer. Sperry Sun downhole recording electronic tool records data on a $\frac{1}{8}$ in wide magnetic tape cassette. Characteristics of all these gauges are given later in Fig. 35.

All these downhole recording electronic gauges take their energy from electrochemical batteries. For the time being, these batteries are one of the main factors limiting the use of the described tools at temperatures higher than 150 °C (300 °F).

3.2. Surface Recording Gauges

Except for one application that will be described in Section 4.4, all surface recording gauges make use of electrical signals. In essence, the sensors are the same as that used in downhole recording instruments. We will describe three typical instruments, each based on a different sensor.

3.2.1. Strain Gauge Based Pressure Transmitter

Figure 13 is the block diagram of a typical pressure transducer. It makes use of a thin film type strain gauge already described in Section 2.2.2.2. In this design, the voltage delivered by the Wheatstone bridge is connected to a voltage controlled oscillator (VCO) which transforms the voltage into a frequency signal. As the voltage is proportional to the pressure exerted on the sensor, the frequency delivered by the VCO will be proportional to the pressure.

As the pressure transmitted is also sensitive to temperature variations, most pressure gauges incorporate a temperature transducer, for example a platinum resistor. The voltage issued from this platinum resistor is

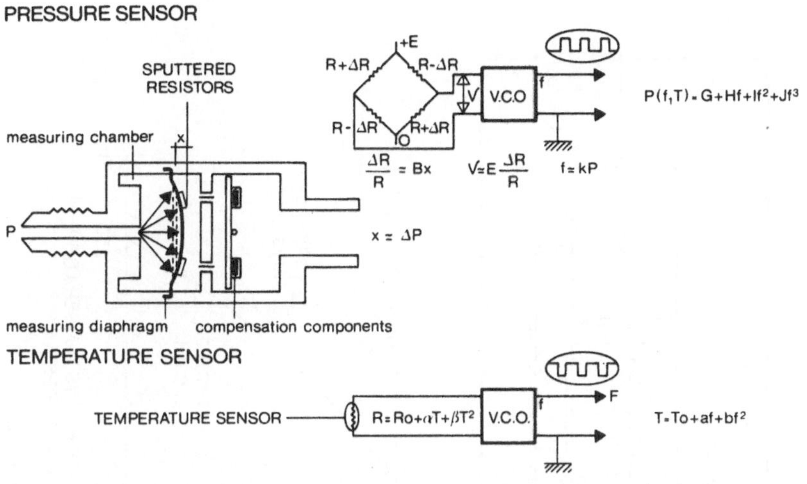

FIG. 13. Block diagram of a downhole pressure and temperature transmitter.

proportional to the temperature, and since this voltage drives a VCO, we end up with a frequency proportional to the temperature.

In a typical make of transmitter, both signals are sent to the surface via an electrical cable: one through positive polarity, the other through negative polarity.

At the surface, the cable is connected to a panel which has three essential functions:

(a) gauge power supply;
(b) signal processing (essentially a frequency counter);
(c) data display, recording, printing.

3.2.2. Quartz Crystal Gauge

Figure 14 is a block diagram of the Hewlett Packard crystal gauge. The gauge contains, in fact, two quartz crystal based resonators. The sensor crystal, subjected to pressure and temperature, is obtained by cutting a cylinder in a specifically oriented quartz crystal block. The active element is a lens-shaped resonator, obtained by careful machining and grinding of this quartz cylinder. Another resonator, restricted to a simple lens and called the reference crystal, is machined with the same crystallographic orientation and preferably in the same crystal; it is subjected to temperature only, not to pressure.

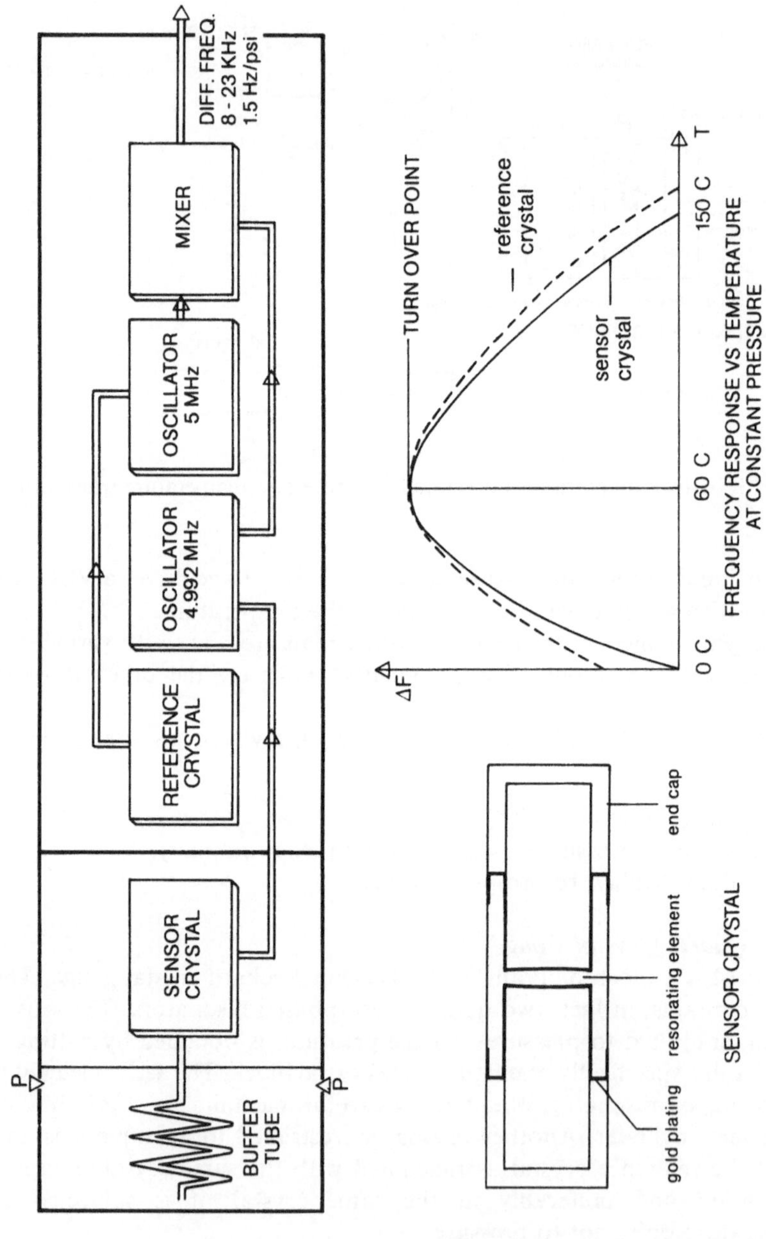

FIG. 14. Hewlett Packard pressure gauge.

The aim in combining these two crystals is to achieve a temperature-independent pressure gauge. For an ideal crystal matching, the two curves of frequency versus temperature (ΔF versus T) shown in Fig. 14 should coincide. In practice, this is never fully achieved, but it can be seen that around 60 °C, variation is minimal.

While the reference crystal is typically resonating at about 5 MHz, the sensor crystal is resonating around 4·992 MHz at 0 psia and about 4·977 MHz at 10 000 psia. Thus the frequency difference varies between 8 and 23 kHz for the full range, or 1·5 Hz/psi. This frequency signal is transmitted to the surface where the processor of the reading panel multiplies it by 72, which results in a sensitivity of 108 Hz/psi.

Since the gauge still remains sensitive to temperature, a temperature transmitter is normally associated with the downhole tool in order to be able to make the necessary corrections.

3.2.3. Capacitive Based Systems
A capacitive gauge has appeared on the market only in the last few years; as already stated, the signal delivered is that of frequency, so that there is no need for a VCO, as in the case with the strain gauge. Compared to the crystal gauge, which also delivers frequencies, it is less expensive. But, so far, too little field experience is available with this gauge.

The performances of all these gauges are given later in Figs 35 and 36.

4. TYPICAL GAUGE APPLICATIONS

4.1. Various Downhole Pressure Gauge Applications
Figure 15 shows a non-exhaustive list of various situations in which downhole gauges may be used. Although this figure is self-explanatory, the following points will now be considered.

There are essentially two types of operations: temporary and permanent. (In this case the gauges used are sometimes called 'implanted gauges'.)

An operation is called 'temporary' when the gauge is run on a cable, or sometimes set inside the tubing (if the gauge is self-contained) for a short-to-medium period of time, typically not exceeding 100 h. During the running in and pulling out operations, the gauge is subjected to shocks and large pressure and temperature variations. For these applications, a gauge which reaches its temperature equilibrium very rapidly will be preferred to others.

An application is called 'permanent' when the gauge is run with the

	DURING DRILLING PHASE	DURING PRODUCTION LIFE				
		TEMPORARY GAUGES			IMPLANTED GAUGES	
		BUILD-UP	WELL TESTING	INTER WELL COMMUNICATION	PERMANENT	RETRIEVABLE
Extreme* conditions	10000 psi 150 °C	10000 psi		150 °C	10000 psi	150 °C
Typical duration of job	8 - 100 hours	1 - 30 days			2 years	1/2 year to 2 years
How the gauge is used	Gauge below DST or in TBG DR or SR	Gauge in tubing DR or SR			Outside tubing SR	In side pocket mandrel,SR
Well condit.	Any	Shut-in	Flowing	Shut-in	Any	
Typical pressure variation	Very high (100 %) of BHP in a few minutes to a few hours	High slow	Medium slow	Small slow	Very slow and low (a few psi to 500 psi/year)	
Typical temperature variation	20 °C	15 °C	10 °C	Nil	Almost Nil except in water injection wells	
Gauge performance requested	High accuracy short time constant	High accuracy and resolution	High accuracy repeatability and resolution no hysteresis	High resolution	High accuracy excellent stability	insensitive to shocks

FIG. 15. Various downhole pressure gauge applications. Key: * may have to be upgraded; DR, downhole recording; SR, surface recording; DST, drill stem testing; TBG, tubing.

tubing, the cable being clamped outside of it; this enables a panel to be connected at any time at the surface for the supply of current to the gauge and reading of the pressure. In this application, the gauge is to remain in the well as long as the completion exists—say, two years typically—and, during this period of time, will be subjected to actual reservoir pressure and temperature. But, except in injection and some gas well cases, pressure and temperature variations to which the gauge is subjected are not very large.

A unique application is represented by the Retrevagage®. The gauge is set in a side-pocket mandrel containing a mechanical locking device and the male part of a connector in liaison with an electrical cable clamped outside the tubing (see Section 4.5.1).

For these permanent applications, one of the prevailing features required from the gauge is the least drift possible under permanent stress.

The temporary applications will now be examined in more detail.

4.2. Measurement During the Drilling Phase

The first opportunity for downhole pressure measurement is offered by the drill stem test (DST). Figure 16 shows the different possibilities.

In a conventional DST, one or more downhole recording gauges are run with the drill string. Figure 16(a) shows a gauge set in the tail pipe of the DST; it will record the reservoir pressure history below the DST valve. In addition, other gauges can be adequately located to check the proper

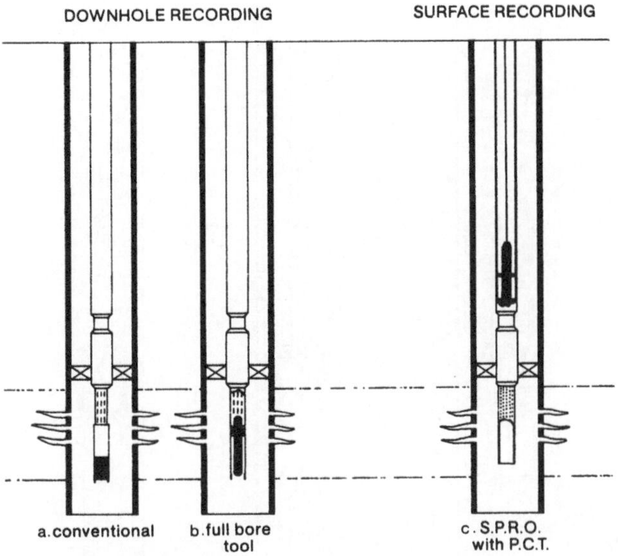

FIG. 16. Downhole pressure measurements with drill stem testing tools.

functioning of the tool or monitor the pressure inside the tubing (i.e. monitoring the water cushion).

In a full-bore tool downhole recording must enable full bore access to be secured when necessary (for example, for stimulation or perforation). Figure 16(b) shows a downhole recording gauge set in a nipple below the DST valve; when full bore access is required, it is easy to retrieve the gauge using standard wire-line techniques.

Figure 16(c) shows the technique called SPRO® which enables pressure to be monitored at the surface, even during the build-up sequence when the downhole valve is closed. A typical application is shown in Fig. 17. The PCT(IM)* tester is run in a well equipped with a pressure/temperature transmitter bearing a male connector at its top. Once the packer has been anchored, a single conductor cable (7/32 in) is run inside the drill string (Fig. 17(b)); the wire-line string incorporates the corresponding female connector which will latch on to the male connector of the pressure transmitter. Then at the surface, pressure can be monitored using an adequate data acquisition system, whatever the sequence: with the PCT closed, as in Fig. 17(b) or with the well flowing, as in Fig. 17(c).

* PCT stands for pressure controlled drill stem testing tool, and refers to the fact that pressurising the annulus operates the tool.

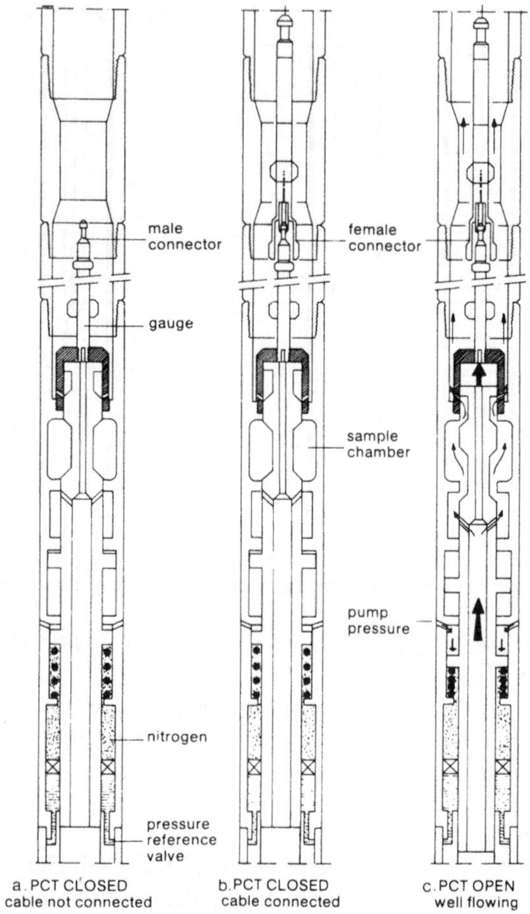

FIG. 17. Surface pressure read-out (SPRO).

Acquisition system capabilities can vary over a large range. In addition, to monitor the test in real time, on-site interpretation facilities are required. An example of a system which includes sophisticated data acquisition and on-site interpretation is the Computest(SM) (Fig. 18). In addition to downhole pressure and temperature transients, Computest also records the flow rates, gas oil ratios or other data coming from well testing equipment installed at the surface. All these data can be used at the well site to monitor and interpret the test in real time; hence, it is possible to verify whether the data are sufficient for a complete interpretation.

FIG. 18. Computest.

4.3. Measurement During the Production Life of the Well

The well is currently equipped with tubing and packer (Fig. 19). If the well is flowing, the access inside the tubing is free and with an appropriate equipment—winch, well head pressure control system—any wire-line tool can be introduced.

Figure 19(a) shows a pressure measurement with the gauge suspended on an electric cable, enabling any real time measurement to be made at the surface.

Figures 19(b) and (c) represent the use of downhole recording gauges; the surface equipment is much lighter than with electric line and surface recording gauges, the drawback being that no real time information is available at the surface. In Fig. 19(b), the gauge remains suspended at the end of the slick line. In Fig. 19(c), the gauge is locked in an appropriate nipple; this solution may be preferred in three cases:

(i) to prevent corrosion of the cable when the well produces a highly corrosive fluid;

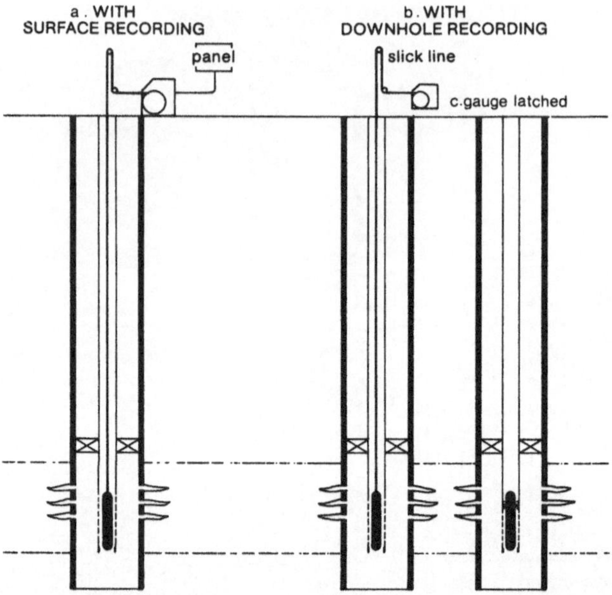

FIG. 19. Downhole pressure measurement with tubing and packer.

(ii) to positively anchor the gauge and avoid it from being lifted up in the case of large producing wells;

(ii) if, for safety reason, the company policy recommends not having any line inside the tubing during a production test (i.e. from a floater).

4.4. Permanent Gauges

In the gauges described in Section 4.3, whenever a downhole pressure measurement is required it is necessary to move a winch to the well site, to rig up the lubricator on the well head and to run the wire-line string into the well.

If for some reason, it is desirable to have frequent access to downhole pressures, these operations may become uneconomical and one must consider the use of permanent gauges. In these gauges the downhole pressure transmitter is mounted on a tubing mandrel (Fig. 20) in the same way as a gas lift valve and is run with the tubing during the well completion. Depending on the type of mandrel, the pressure can be measured from either inside the tubing or the annulus. Connection from the bottom to the

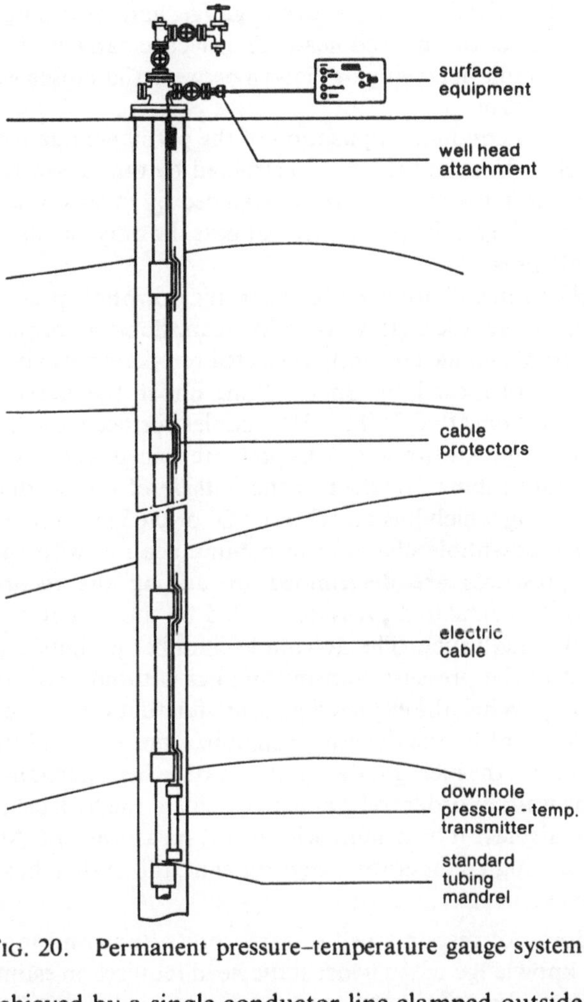

FIG. 20. Permanent pressure–temperature gauge system.

surface is achieved by a single conductor line clamped outside the tubing during its running into the well. Adequate well head attachment enables readings to be made on the spot whenever desired or permanent connection to a control room.

Permanent gauge installations are also recommended on subsea well heads thus enabling downhole pressure monitoring to be made without requiring an expensive and sometimes hazardous subsea wire-line operation. Specific technology has been developed to achieve sea floor connection between the tubing hanger, run with the tubing and containing the connection to the downhole pressure gauge, and the tubing head set via the work-over riser on the well head. The electric cable is clamped to the other service bundle achieving the liaison between the subsea well head and the service platform.

Finally, a very common application of the permanent gauge is its use in pumping wells when the tubing is obstructed by the sucker rods or by an electric cable as is the case for wells produced by means of a submersible electrical pump. Here, the permanent gauge is the only means of measuring the downhole pressure.

One application still not open to electronic downhole pressure gauges is measurement in geothermal wells or in steam flooding projects, since no reliable electronic gauge currently exists for temperatures above 400 °F. A unique design proposed by Sperry Sun, under the name of pressure transmission system (P-T-S) (Fig. 21), enables surface monitoring without using any electronics downhole. This pressure transmission system consists of a 3/32 in OD tubing attached to the outside of the production string. This small tubing which has an ID of 0·026 or 0·054 in connects a surface recorder to a downhole chamber in communication with the well fluid. Downhole pressures are determined by adding the recorded surface pressures to the calculated pressure exerted by the weight of an inert gas column in the small tube. The downhole chamber permits expansion and compression of the pressure transmitting gas without entry of well fluids into the tubing. When the expected pressure fluctuations are too large to be taken into account by the downhole chamber, an open-end tubing can be used. In this case, the inert gas (helium) may have been partially replaced by gas from the well; in order to be able to calculate the hydrostatic head it is necessary to displace it by helium prior to any measurement. Although very attractive for a high temperature environment, this system has, essentially, two drawbacks:

(i) it does not respond well to rapid pressure variation;
(ii) the knowledge of the hydrostatic head requires an estimation of the temperature distribution versus depth.

4.5. Retrevagage (Fig. 22)

A permanent pressure gauge is continuously exposed to downhole pressure and temperature. If a stimulation job appears to be necessary, the gauge will be exposed to abnormal pressures or to stimulation fluids, and may well cease functioning. The only way to replace the gauge for repair or to recalibrate it, is to pull the tubing.

A unique design, called 'Retrevagage', enables the replacement of the

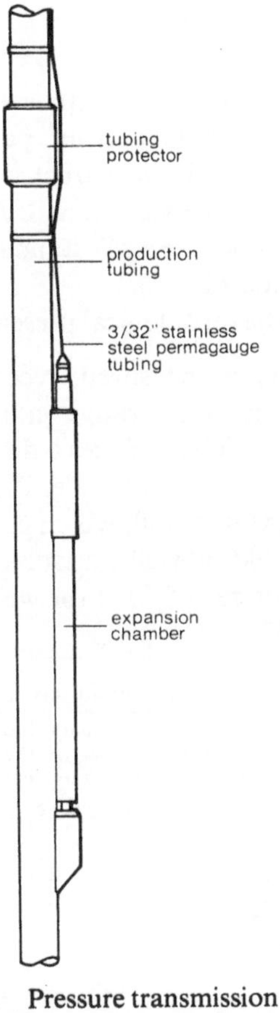

FIG. 21. Pressure transmission system.

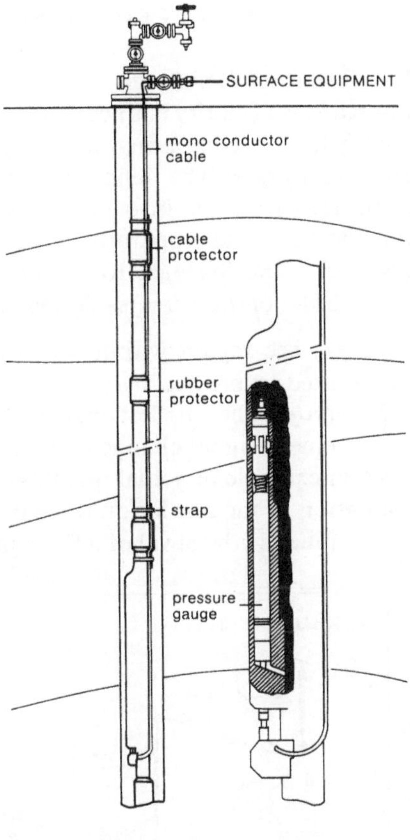

FIG. 22. Retrevagage.

gauge to be made without pulling the tubing. The retrievable gauge is placed inside a side-pocket mandrel run with the tubing string. The side pocket has at its upper part a mechanical lock and at its lower part a male connector in contact with the cable clamped outside the tubing as for the case of the standard permanent gauge. Should it become advisable to replace the gauge, e.g. for recalibration, or to remove it temporarily (in the case of a stimulation job, to avoid its damage or destruction), the gauge can be retrieved using conventional slick line techniques: a kickover tool is run to retrieve the gauge and a dummy is set inside the pocket with a view to protecting the male connector from being damaged by acid or fracturing fluids. After the job the gauge can be rerun in the mandrel.

5. PERFORMANCES OF THE GAUGES

5.1. Review of Definitions

For reasons of clarity it is necessary to define the terms currently used when qualifying or comparing metrological instruments and in particular pressure gauges. The definitions given here have been extracted, where applicable, from the Instrument Society of America standards ISA-S51,1 and the Scientific Apparatus Makers' Association (SAM) standard No. PMC 20-1 and SAMA's RC 20-11-1964 standards.

The order of the terms is selected in view of the most logical presentation.

Range: the region within which a quantity is measured, received or transmitted, expressed by stating the lower and upper range values.

Resolution: the smallest change of variable which produces a detectable measurement signal change expressed in per cent of full scale. Figure 23 gives an example of actual measurements made in a well, which enable an evaluation of the resolution to be made; this old oil well was subjected to discontinued withdrawal of a few cubic centimetres of oil at the well head;

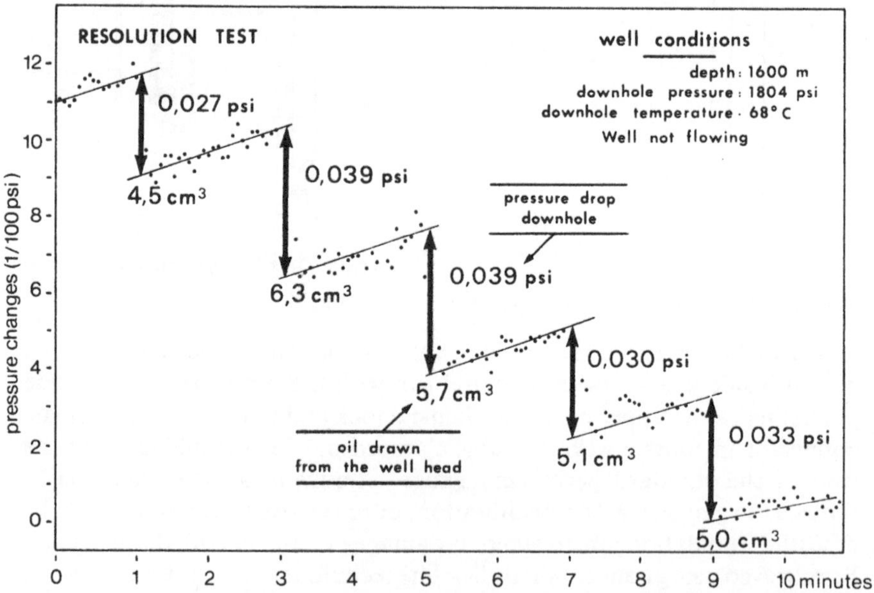

FIG. 23. Resolution obtained with a 2500 psi gauge. A downhole pressure drop of 0·027 psi is registered after 4·5 cm^3 of oil has been drawn from the well. This resolution test shows the high sensitivity of this pressure gauge.

between two withdrawals, the rising of the pressure curves shows the recompression of the reservoir.

Hysteresis (Fig. 24): the difference of the signal measured for a given variable value when approached first from a zero load and then from full scale.

$$\text{Per cent hysteresis} = 100A/B$$

where A = maximum difference in signal between the increasing and decreasing run of the initial calibration and B = signal at full scale.

Drift: an undesired change in output over a period of time. Drift is usually expressed as the change in output over a specified time with fixed input, and operating conditions.

Repeatability (Fig. 25) (precision): the closeness of agreement among a number of consecutive measurements of the output for the same value of the input under the same operating conditions, approaching from the same direction, for full range traverses.

Conformity (of a curve) (Fig. 26): the closeness to which it approximates a specified functional curve (e.g. logarithmic, parabolic, cubic, etc.).

Linearity (Fig. 27): the closeness to which a curve approximates a straight line.

Accuracy (Fig. 28): the limits within which the stated value of a variable property might vary relative to its true value. Generally expressed in per cent full scale. (Example: $\pm 1\,\%$ full scale accuracy of 0–10 000 psi pressure gauge means that the accuracy is $\pm$ 100 psi both at a pressure of 4000 psi or at 8000 psi.) Accuracy is affected by:

—repeatability;
—errors induced during calibration;
—system induced errors (clocks, frequency meter, etc.);
—conformity;
—hysteresis.

In practice, we have to accept the first three sources of errors. We can control and minimise the influence of the last two by:

—using an adequate data exploitation programme to minimise conformity error;
—using adequate calibration sequences or procedures in order to reproduce the same sense of variation of the parameters to which the tool will be exposed in an actual case.

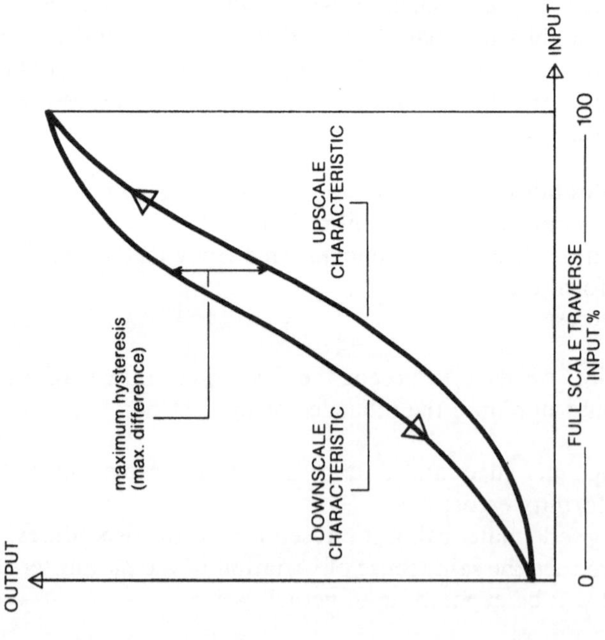

FIG. 24. Hysteresis.

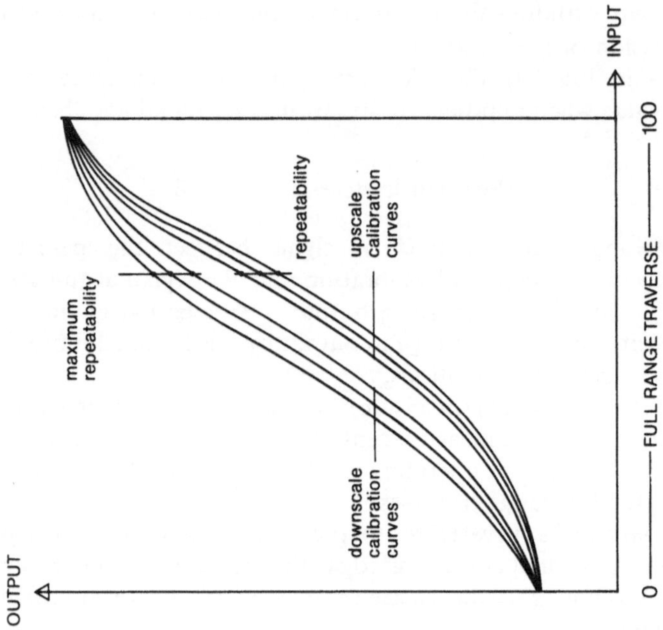

FIG. 25. Repeatability.

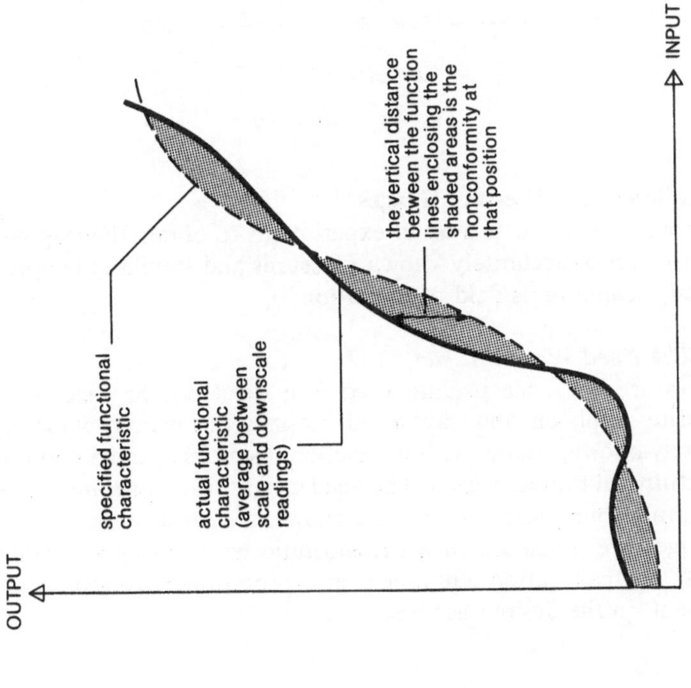

FIG. 26. Conformity.

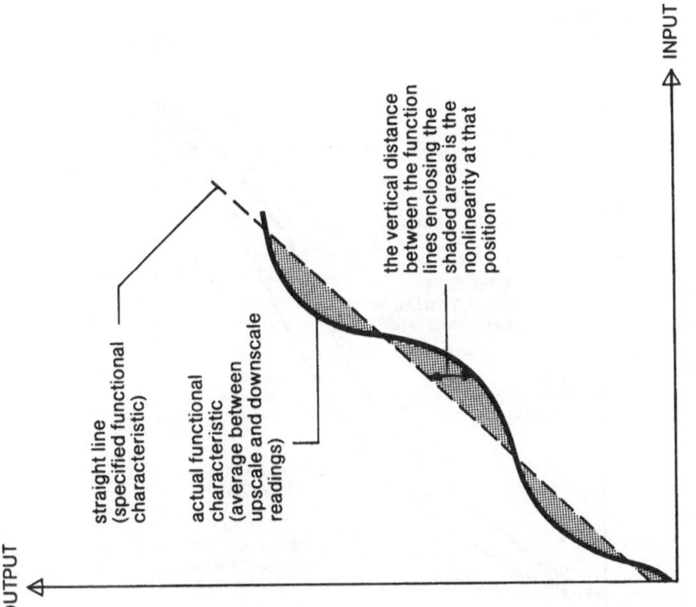

FIG. 27. Linearity.

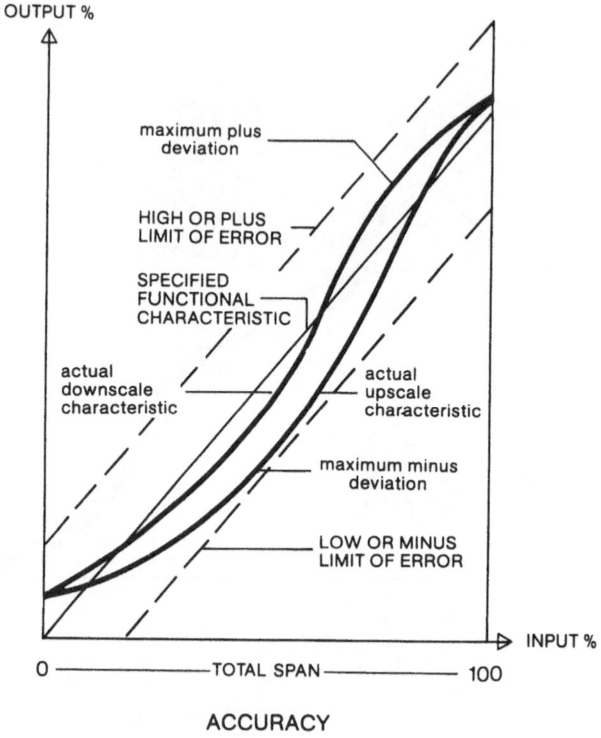

FIG. 28. Accuracy.

5.2. Calibration of Pressure Gauges

We denote calibration as a set of experiments to obtain the response of a sensor to a set of accurately known pressures and stabilised temperatures uniformly scanning its field of operation.

5.2.1. The Dead Weight Tester

The pressure reference usually used is a dead weight tester (Fig. 29) comprising a piston and calibrated masses; the piston, whose area is accurately known, rotates without friction inside its cylinder and is loaded by carefully calibrated masses. The dead weight tester operates at ambient temperature, but because downhole gauges are sensitive to temperature, the gauges are immersed in a thermostatic bath during the calibration process. The calibration equipment and procedure involved are directly dependent on the desired accuracy.

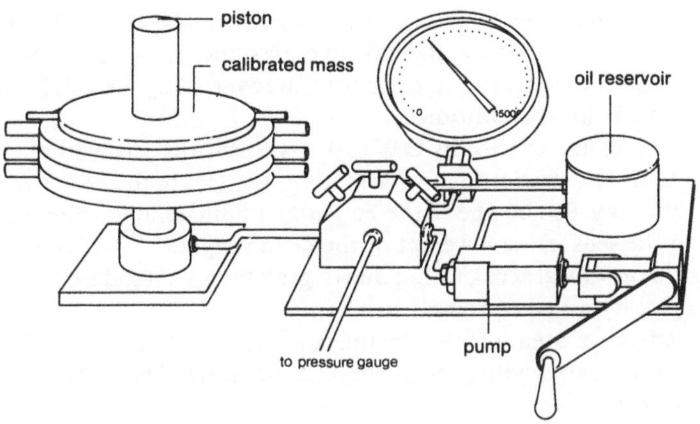

FIG. 29. Dead weight tester.

5.2.1.1. High accuracy gauge calibration. In addition to the obligatory use of master class dead weight testers for high accuracy calibration, the temperature of the room, the atmospheric pressure and local gravity corrections must also be taken into account whenever a calibration must provide a pressure reference with an accuracy of $\pm 1 \times 10^{-4}$ of the reading.

5.2.1.2. Gravity effect. The most important factor in dead weight tester (DWT) calibration accuracy is an exact value for the earth's acceleration due to gravity, g. All mass units used by DWT manufacturers are converted into weights for a standard acceleration, i.e.

$$g(\text{standard}) = 980{\cdot}665\ \text{cm/s}^2 = 980\,665\ \text{milligal}$$
$$= 32{\cdot}174\,05\ \text{ft/s}^2$$

The local acceleration due to gravity (in ft/sec^2) is calculated from the 1980 'International formula':

$$g = 32{\cdot}088\,2\{1 + (5{\cdot}288\,4 \times 10^{-3})\sin^2 \emptyset - (5{\cdot}9 \times 10^{-6})\sin^2 2\emptyset\}$$

where $\emptyset$ is the latitude in degrees. In addition, there is a local anomaly (the Bouguer anomaly), a correction for the altitude and also a correction for the well depth.

Since variations in g can be as great as $\pm 0{\cdot}3\,\%$ from the world standard value, with positive variations in the polar vicinity and negative variations in equatorial regions, it is necessary to take this acceleration due to gravity effect into account.

5.2.1.3. Calculation of pressure generated by the DWT. To meet the required level of accuracy, the effective absolute pressure exerted on the sensor without weights (p_e) and the pressure exerted by the weights must be calculated for local conditions.

Let p = pressure exerted by DWT; p_e = pressure exerted without weights at standard acceleration due to gravity g_s (p_e is due to the weight of the piston but may also be due to the height h of fluid column from the weight stand to the sensor); w = weight of the dead weights; A = effective area of the piston; g = local acceleration due to gravity; g_s = standard acceleration due to gravity = 980·665 cm/s^2.

The effective area of the piston of the DWT determined by the manufacturer varies with pressure and temperature. The empirical relation is of the form:

$$A = A_0(1 + bp)\{1 + c(T - T_0)\}$$

where A_0 = effective area at zero pressure gauge and at reference temperature T_0; b = fractional change in area per unit pressure due to elastic distortion; c = coefficient of superficial expansion for the combination of materials used for the piston and cylinder; T = temperature. Values of b and c are supplied by the manufacturer. Typical values of b and c are

$$b \simeq -5 \times 10^{-8} \text{ per psi}$$

$$c \simeq 1{\cdot}5 \times 10^{-5} \text{ per °C}$$

Therefore, at 10 000 psi, the pressure correction is around 0·05 % or 5 psi for the change in area due to pressure. A temperature difference $(T - T_0) = 10$ °C introduces a correction of $1{\cdot}5 \times 10^{-4}$ of the reading or 1·5 psi at 10 000 psi.

It is obviously interesting to have the calibration laboratory built in a special room air-conditioned at 20 ± 0·5 °C and 40–50 % relative humidity.

The pressure applied to the gauge is:

$$p = (p_e + w/A)g/g_s$$

Theoretical analyses about uncertainties in dead weight tester pressures have been made by manufacturers and by users. They agree on 1 psi of uncertainty at the 10 000 psi level, if all corrections and the utmost care are put into the calibrating experiments.

5.2.2. Calibration Bath

Because of the high sensitivity of pressure gauges to temperature variations it is essential that the gauge is entirely immersed in a temperature-controlled bath during its calibration process; some gauges are, in addition, very sensitive to the temperature gradient along the instrument itself.

For these reasons high-quality calibration baths usually have a high thermal inertia—which requires long heating or cooling times—and are designed with a circulation of oil (by pump or simply by natural convection) in order to maintain the gradient within a few hundredths of a degree Celcius. The temperature of the bath is measured with a platinum resistor enabling accuracy better than 0·1 °C to be guaranteed.

5.2.3. Mathematics in Retrieving Pressure from Frequency

The rate and the accuracy of pressure data collection by modern pressure sensors makes it expedient to convert the frequency and the temperature into the absolute pressure measurement by calculation, or by means of electronic computers.

From the set of calibrating experiments and by least-squares fitting we derive a polynomial with two arguments:

$$\begin{aligned} p &= G_0 + G_1 T + G_2 T^2 + G_3 T^3 + \cdots \\ &\quad + H_0 f + H_1 f T + H_2 f^2 T^2 + \cdots \\ &\quad + I_0 f^2 + \cdots \\ &= P(f, T) \end{aligned}$$

Usually eight coefficients are sufficient to fit the calibration over the entire range of operating conditions for a strain gauge sensor within its physical accuracy.

The following optimisation (or regression) efficient routine was devised, based on Legendre polynomials.

By linear substitutions we change the operating range of the physical variables p, f and T into the range -1, $+1$, i.e.

$$p(\text{new}) = \frac{2p - p_{\min} - p_{\max}}{p_{\max} - p_{\min}}$$

and similar substitutions can be made for f and T.

After these substitutions the simplest approximation of the new function P is the identical function in the interval $(-1, +1)$.

$$p(\text{new}) = f(\text{new})$$

corresponding to a physical function limited to

$$p = G_0 + H_0 f$$

In what follows, we write in terms of new (or reduced) p, f and T.

The best fit is characterised by the minimum of the 'least-squares principle' expression:

$$\min \frac{1}{N} \sum_{i=1}^{i=N} \{p_i - P(f_i, T_i)\}^2$$

The concept of orthogonal functions appears when we consider the analogy between the preceding 'least-squares principle' expression for a uniformly distributed (in f, T) calibration and the following integral:

$$E = \int_{-1}^{+1} \int_{-1}^{+1} (p - \alpha_1 P_1 - \alpha_2 P_2 - \alpha_3 P_3 - \cdots)^2 \, \mathrm{d}p \, \mathrm{d}T$$

which is the limit of the least-squares principle for a set of p, f, T tending to an infinity of uniform density in the rectangular domain of integration, itself representative of the operating range of the sensor.

To minimise this integral, i.e. to determine the set of coefficients making up the optimal P function out of the set of the component functions P_i, we have to solve the following linear system:

$$\frac{1}{2} \frac{\partial E}{\partial \alpha_i} = 0$$

Let us select for component functions the products of normalised Legendre polynomials in p and T:

$$P_i = L_m(f) L_n(T)$$

Noting that the orthonormality properties of Legendre polynomials:

$$\int_{-1}^{+1} L_m(x) L_n(x) \, \mathrm{d}x = 0 \qquad \text{if } m \neq n$$
$$= 1 \qquad \text{if } m = n$$

we find that the matrix of the above-mentioned linear system becomes the unit matrix, which was the aim of our selection, and the solution of this system is

$$\alpha_i = \int_{-1}^{+1} \int_{-1}^{+1} p L_m(f) L_n(T) \, \mathrm{d}p \, \mathrm{d}T$$

As a consequence of the orthogonality properties, each coefficient brings an independent contribution to the minimisation of the E integral.

We have programmed a computer to obtain, over the finite set of experimental measurements, the α coefficients by order of physical significance among the 25 products of the first 5 Legendre polynomials in reduced f and T. The computer feature known as 'double precision' is necessary in this calculation.

The number of significant terms is normally found to be eight, after which we obtain an irreducible random character of the residual errors. The method indicates by itself the quality of the calibration and of the gauge.

The program after the determination of the function P, goes back to the true physical scale for p, f and T and sorts, by degree, the various terms generated by the Legendre polynomials. We obtain the formula:

$$p = G_0 + G_1 T + G_2 T^2 + G_3 T^3 + H_0 f + H_1 f T + H_2 f T^2 + I_0 f^2$$

Use of Legendre polynomials in this case is a way to generate a sound set of least-squares fit equations linked with a nearly diagonal matrix.

5.3. Ageing of Gauges

Ageing is the phenomenon according to which the characteristics of all materials change with time: wood dries, copper tends to be covered by verdigris. Steel used to make pressure gauges, and even quartz, follow the same law. Although, for some materials, ageing consequences result in variations of a second-order magnitude, whenever accurate measurements have to be made, ageing has to be taken into account.

One good method of monitoring the ageing of pressure gauges is to calibrate them from time to time, and compare the evolution of the different calibration coefficients and overall results.

Ageing appears as a drift of the response to pressure versus time; the drift is specific to each gauge and no average value can be given. One can try to give general hints as far as the magnitude of the drift is concerned:

—the drift of a gauge subjected to ambient temperature and atmospheric pressure is almost nil;

—the drift is lower for quartz gauges than for strain gauges;

—the drift starts to be noticeable when the gauge is exposed to pressure and temperature variations over long periods of time;

—the longer the time, the higher the pressure as compared to full range, and the higher the temperature, the more important the observed drift;

—drift tends to increase more rapidly during the first period under stress, then tends to level off and to become asymptotic after a couple of months.

Only a few engineers have studied this phenomenon extensively and no general rules can be deduced so far from these studies.

5.4. Thermal Behaviour of Pressure Gauges

All available pressure gauges are sensitive to both pressure and temperature variations. Since the petroleum engineer is essentially interested in the pressure variation of the reservoir, he must make certain that the pressure recorded has been corrected for the temperature to which the gauges have been exposed. This is the reason why pressure gauges have to be calibrated over all the range of temperature they may have to be exposed to in the future.

All manufacturers have devoted great efforts to compensate their gauges for the temperature range for which they are sold. Thus, the Bourdon type used in the Amerada gauge are made of Ni Span C which is almost insensitive to temperature, and the strain gauges contain built-in resistors which aim to compensate for temperature effects.

In fact, these compensations never cover the total range of application and in addition do not allow the present required standards to be attained.

Below it will be explained how one can take into account the temperature factor but, first, two cases must be distinguished: stabilised temperature and transient temperature conditions.

5.4.1. Stabilised Temperature Conditions

In a production well, shut-in for a few hours or days, the downhole temperature is constant, whereas in an injection well the temperature might rise slightly until the current reservoir temperature is reached.

During a well test operation, the temperature may change by a few degrees Fahrenheit during the shut-in and flowing period, and in the case of a gas well, variations can be a few tens of degrees Fahrenheit. But once a given phase—i.e. a given flowrate—has been set, the temperature may stabilise.

From now on, all pressure calculations must be corrected for the actual temperature of the electronics, and since we assume that the conditions are stabilised, the temperature of the electronics will be that of the well at the depth of the pressure gauge.

The temperature is measured; knowing the calibration coefficient the

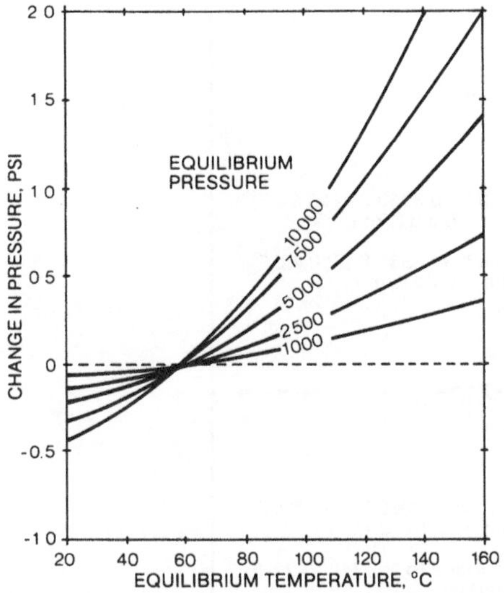

FIG. 30. Hewlett Packard gauge: pressure change due to a 1 °C temperature change under stabilised conditions.

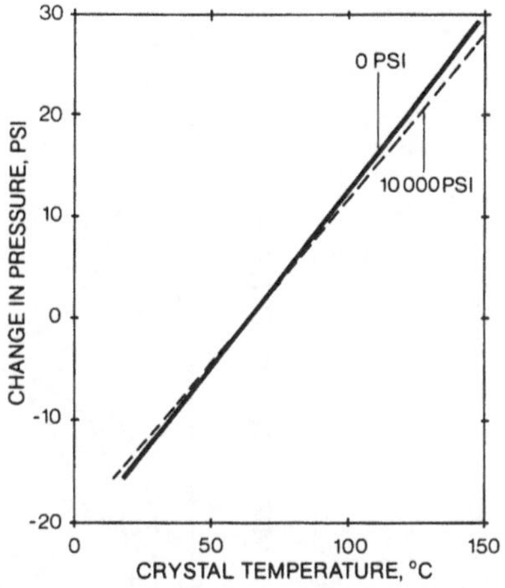

FIG. 31. Hewlett Packard gauge: pressure difference due to a 1 °C temperature difference between the sensor and the reference crystal.

temperature value is introduced into the computer and the pressure calculation is then corrected for any temperature effect.

The effect of the temperature on a measured pressure depends on the type of gauge used and on the level of pressure. For a strain gauge, typical values are 2–6 psi per °C. For a Hewlett Packard quartz gauge the behaviour can be seen in Fig. 30. One recognises the turn-over point already mentioned and notes that the pressure error increases with both average pressure and temperature.

5.4.2. Transient Temperature Conditions

In many circumstances, pressure gauges are submitted to temperature variations and pressure readings are made while thermal equilibrium has not yet been attained throughout the pressure gauge transducer. Laboratory tests have shown that when imposing a 1 °C temperature difference between the two crystals of a Hewlett Packard quartz gauge, the error effect on pressure readings can be in excess of 20 psi (Fig. 31). Although such a high temperature difference may not exist in practice, actual field measurements have shown the reality of this phenomenon.

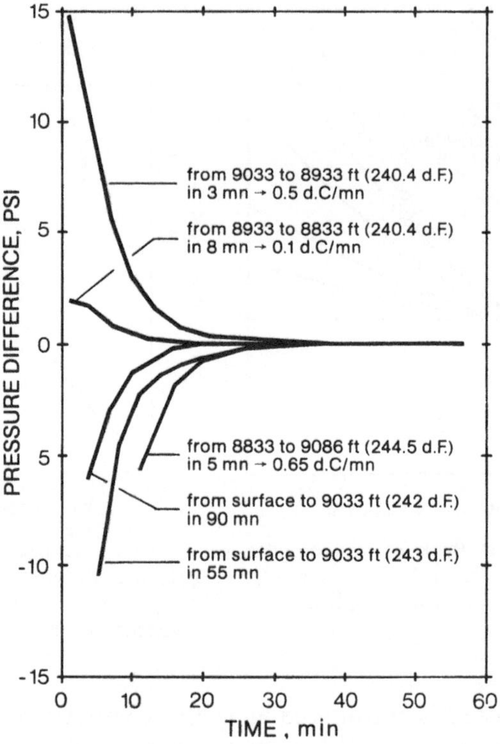

FIG. 32. Hewlett Packard gauge: difference between true and recorded pressures during pressure gradient measurements.

Figure 32 taken from field examples illustrates the order of magnitude observed in actual cases and the time required to attain stabilization.

Laboratory studies have been made to understand this process and the main results are shown in Fig. 33. The work concerned three gauges subjected to a very rapid temperature jump of +35 °C (from 20 °C in air to 55 °C in oil). The reactions of these gauges have been plotted versus time and gave the following results:

1. A standard quartz gauge gave an overshoot of about 80 psi and it took about 45 min to get an acceptable reading.
2. For a strain gauge, the overshot was about 20 psi but the time to reach equilibrium was only slightly shorter.
3. The third gauge concerned a repackaged quartz gauge. The behaviour is better by far and an acceptable reading can be obtained after about 10 min.

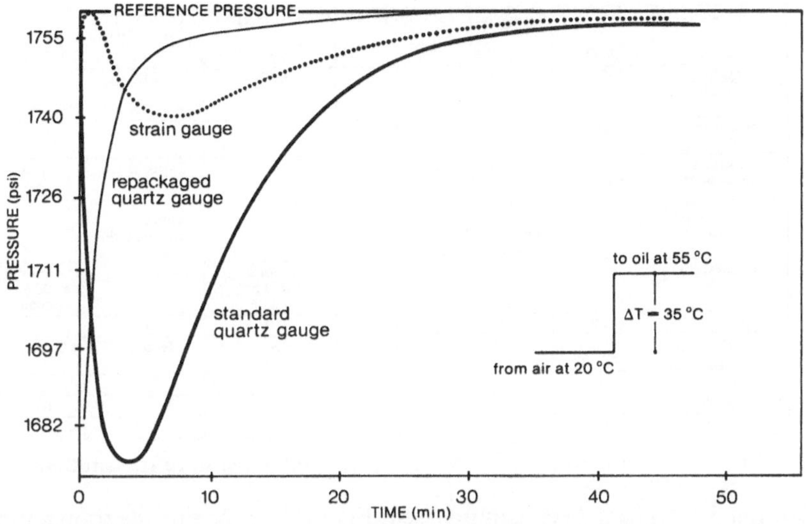

FIG. 33. Thermal transients at constant pressure—simulation of oil well conditions.

5.5. Pressure Transient at Constant Temperature

For completeness, it should be mentioned that rapid pressure variations also generate observable transient phenomena as shown in Fig. 34. As shown, the order of magnitude of this phenomenon is far smaller than that observed in thermal transients.

5.6. Comparison of Six Electronic Pressure Gauges

In Figs 35 and 36 the main characteristics of currently available electronic pressure gauges (characteristics of mechanical gauges have been given in

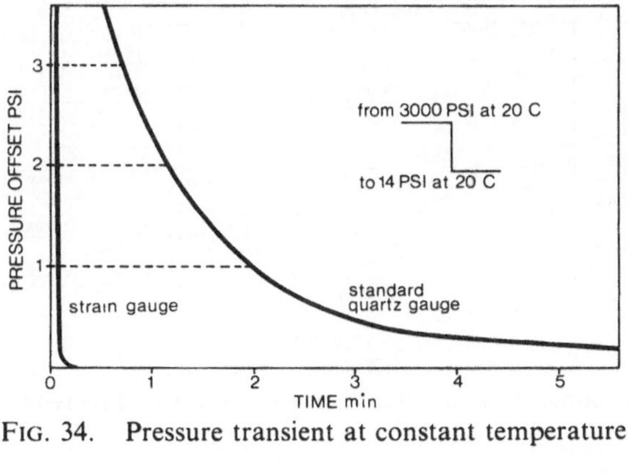

FIG. 34. Pressure transient at constant temperature.

Manufacturer	FLOPETROL	HEWLETT PACKARD	SPERRY SUN	PAROSCIENTIFIC	GEOPHYSICAL RESEARCH CO	LEUTERT
Name	DPTT	2811 B	–	75 K	EPG 512	DTE 80
Pressure range psi	10 000	11 000	15 000	5 000	10 000	8 700 (14 500 in preparation)
Temperature range °C	150	150	150	107	150	150
Pressure transducer	Thin film strain gauge	Quartz crystal	Bourdon tube	Quartz crystal	Capacitance	Piezo resistive
Accuracy (in psi) for a 10 000 psi gauge	4	0,5 psi up to 2 000 psi Above: 0.025 % of reading	5		4 - 10 Depending on the range	
Repeatability				4 (for a 5 000 psi gauge		8.7 (for the 8 700 psi gauge
Resolution (in psi)	0.02	0.01	0.5	0.01	0.2	1
Temperature transducer	Platinum Resistor	No	No info.	No	Resistor	I.C. circuit
Accuracy (°C)	0.3				0.7	
Repeatability (°C)						0.3
Resolution (°C)	0.03				0.03	

FIG. 35. Comparison of downhole pressure/temperature transmitters.

Section 3.1.1.) have been summarised. Figure 35 concerns the transmitters themselves. Data have been taken from manufacturers' leaflets and in many cases have been checked by Flopetrol. When accuracy figures are given, it has to be understood that they imply that temperature is known to within an accuracy of 1 °C or better.

Figure 36 refers to downhole recording gauges; when comparing, for a given manufacturer, the same pressure sensor used in surface recording (Fig. 35) or in downhole recording, it can be seen that the accuracy will be slightly lower in the second case. The reason is that when downhole recording is considered, the crystal driving the frequency counter is also exposed to high temperature, whereas in the surface recording system, the frequency counter is at ambient temperature.

Make	FLOPETROL			LYNES		SPERRY-SUN
Name	SSDR/DPTT	SSDR/HP	SSDP	DMR 312 PTS	DMR 314 PTX	
Type	Strain gauge	Quartz Crystal	Strain gauge or Quartz crystal	Strain gauge	Quartz crystal	
Pressure range	10 000	11 000	as SSDR	10 000	5 000	
Temperature range (°C)	150		150	125	107	150
Data storage capacity	5120		5120 but DRA*	1024		
Storage memory	E. PROMS		E PROMS + EE PROMS	EE. PROMS		MAG TAPE
Minimum sampling rate	30 sec.		1 sec.	3 75 sec.		15 sec.
Delayed start	NO		0-64 hours	0 - 7 hours		0 - 17 hours
Accuracy for 10 000 psi gauge	± 5 psi	+ 0.035 % for reading above 2 000 psi	as SSDR	± 10 psi	–	± 5
Accuracy for 5 000 psi gauge	± 2 psi			± 5 psi	± 2 5 psi	± 2.5
Resolution (in psi)	0.02		0.02	2.5	1 for 5 000 psi gauge	0.5
Hysteresis	< 3 psi	Almost nil	as SSDR	?	?	?

FIG. 36. Downhole recording electronic gauges. DRA, data reduction algorithm.

6. SUGGESTIONS FOR BEST FIELD USE OF HIGH ACCURACY PRESSURE GAUGES

No ideal gauge exists, so far. Pressure gauges have the drawback of being also, to a certain extent, thermometers! They drift, are subject to hysteresis and are influenced by thermal transients.

Considerable work is being done by many research centres to improve the gauges but, for the time being, we have to live with what is available. A good understanding of the behaviour of the gauges, and of their defaults, is a must when planning a pressure measurement campaign. Only with long practice can the best advantages from a gauge be obtained. Three important hints to consider are:

—Be sure that any gauge used has been recently calibrated.

—Monitor the evolution of the gauge by comparing the successive calibrations to which the gauge has been subjected.

—Whenever possible, programme your operation in such a way as to minimise the thermal transient and hysteresis effects.

Chapter 5

INTERPRETATION OF TRANSIENT WELL TEST DATA

A. C. GRINGARTEN

Scientific Software—Intercomp, Denver, Colorado, USA

1. INTRODUCTION

1.1. Purpose of Well Testing

Petroleum engineers in oil companies must answer two fundamental questions:

1. What is the most efficient way to produce the reservoir?
2. What is the most efficient way to produce each well in the reservoir?

An answer to the first question requires the establishment of a reservoir development strategy, based on two types of models:

—an economic model
—a reservoir model

The economic model takes into account present and projected economic information, such as: the price of oil and gas, interest rates, return on investment, etc.

The reservoir model is used to predict the physical behaviour of the field, in terms of rate of production, and fluid recovery for different operating scenarios, including various development patterns, primary production and secondary or tertiary recovery processes.

In order to perform realistic predictions, the reservoir model should represent the behaviour of the actual reservoir as closely as possible. It is built using available information from geology, geophysics and measurements performed in wells. The reservoir model is never definitive, but must continuously be adjusted as new information becomes available when additional wells are drilled.

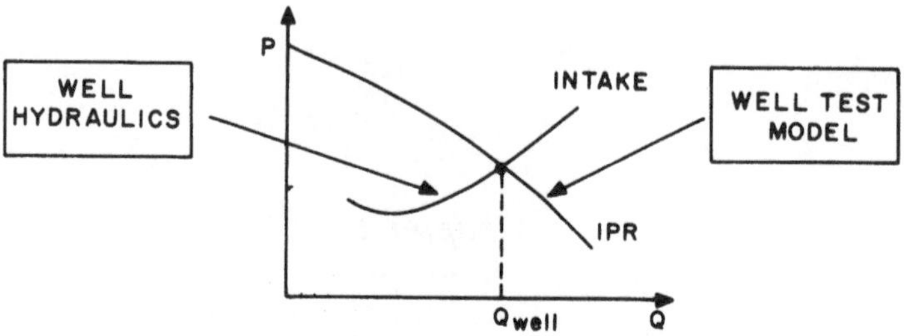

FIG. 1. Well performance evaluation.

Well testing is one of the many sources of information for building a reservoir model. In most cases, input to reservoir modelling is the only purpose for testing a well. Yet, in general, only a fraction of information available in a test is used in the reservoir model. The rest of the information goes largely undetected. Another important use of well testing results, however, and certainly the one that impacts most immediately on oil company revenues, is the design of the well completion to ensure optimal production of each well in the reservoir.

The use of well test results in completion design is illustrated in Fig. 1, which is a graph defining the well capacity. The top curve labelled IPR (Inflow Performance Curve), represents the pressure–rate relationship in the reservoir. The lower curve, called INTAKE, represents the pressure versus rate relationship in the well itself from the bottom to the surface facilities. The well can only produce at a specific rate which is obtained as the intersection of the IPR curve with the INTAKE curve. Finding ways to increase the well production requires the evaluation of the various ways that can lead to the modification of either the IPR or the INTAKE curve. The manner in which either curve behaves determines whether the well is tubing or reservoir dominated. There is no advantage, for instance, in fracing a well if it is tubing dominated, whereas this would be a viable option if the well were reservoir dominated.

The INTAKE curve is computed using well-established flow correlations for the various elements of the production system (valves, tubing, pipe, separator, etc.), and is usually defined with little ambiguity, within accepted industry standards. The IPR curve, on the other hand, is a strong function of the flowing characteristics of the reservoir (permeability, damage and behaviour) that can only be obtained by testing the well. Information obtained from other sources such as wireline logs or cores is usually inadequate and could result in completion decisions that could adversely affect the economics of operating the well. For instance, a well completion

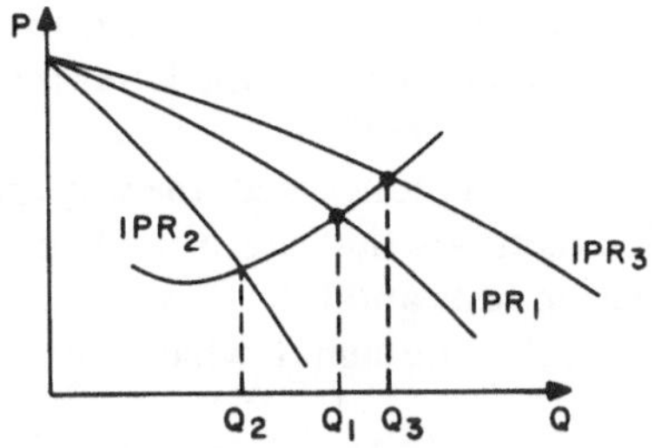

FIG. 2. Effect of reservoir behaviour on well performance.

based on the IPR curve number 1 in Fig. 2 for a rate Q1, would only produce a rate Q2 much less than Q1 if the actual IPR curve of the well were the curve number 2 instead of the curve number 1, whereas the rate could be much higher if the analysis of the test indicates the possibility of improving the formation, thus resulting in the IPR curve number 3.

Well testing is therefore mandatory for the engineering of reservoir performance as well as for the engineering of well performance. In this chapter, we describe a general and systematic approach to the interpretation of well test data.

1.2. Inverse versus Direct Problem[1]

Well testing provides an indirect means of acquiring knowledge on underground formations. Direct measurements in a reservoir are only possible on fluid samples, to determine the nature and composition of the fluid, and on cores. Core measurements, however, only provide static information, which is not adequate for defining the production potential of the reservoir. This can only be obtained indirectly, by studying the behaviour of the reservoir under dynamic conditions. The objective of well testing is to create a well-defined dynamic environment.

Well testing is a typical signal analysis problem, schematically described as

$$I \rightarrow S \rightarrow O$$

where a known input signal I is applied to an unknown system S, resulting in a response of the system, the output signal O.

In well testing, the input signal is usually a step function in rate, created when a flowing well is closed, or a shut-in well is open to production. The output signal is the change in pressure at the bottom of the well created by the change in rate. The rate signal can be created at the surface, by shutting or opening the master valve at the well head, or at the bottom of the well, by using a special downhole shut-in valve. The former is more commonly used in wells already in production, whereas the latter is standard practice after drilling (drill stem testing or DST).

As far as well test analysis is concerned, the way the rate signal is created is not important. The methods that are described hereafter are perfectly valid for both production tests and DSTs. What is most important for the analysis is the quality of the rate signal, which must be of the proper shape and proper duration.

The objective of well test analysis is to identify and describe the unknown system S, by extracting all the information available in the input and output signals. This is known in mathematics as the inverse problem, symbolically represented by

$$S = O/I$$

Solving the inverse problem implies an identification or pattern recognition process. Interpretation provides a model whose behaviour is identical to the behaviour of the actual reservoir. The solution of the inverse problem, however, is usually not unique. It may be possible to find several reservoir configurations that would yield similar responses to a given input signal. In general, the number of alternative solutions decreases as the number and the range of output signal measurements increase.

The inverse problem is different from the direct problem:

$$O = I \times S$$

which assumes a specific model S, and adjusts its main parameters until the computer response O to the actual input signal I is as close as possible to the measured output signal. In well test interpretation, the direct problem is solved once the interpretation model has been identified.

Attempting to solve the direct problem before the inverse problem has been solved is likely to yield erroneous results in interpretation, because it involves the a priori selection of an interpretation model based on static information not on test behaviour. Unfortunately, the ability to describe the geology and the geometry of the actual reservoir does not guarantee that an adequate model for interpreting tests in that reservoir can be obtained. In other words, the test behaviour cannot be predicted from static reservoir description.

1.3. The Concept of Interpretation Models

This point is fundamental to the understanding of the concept of interpretation models. The interpretation model is in fact a transfer function, and only describes a specific response of the system to a specific signal. If the nature of the input or output signals is changed (e.g. temperature instead of pressure), so is the interpretation model.

This point is illustrated in Fig. 3[2] which presents three levels of information about the reservoir: the first level includes the measurements, or raw data, obtained from geology, geophysics, logs, or well testing. The other two levels represent two different types of models. The first type of model comes from interpretation of measurements: geological data yield a geological model, log data yield a log model, and well test data yield a well test model. Because the geologist, the log analyst, and the well test interpreter see different aspects of reservoir behaviour, the models they identify are independent, and have little in common: in fact, a model derived by one has little direct use for the other two. For instance, the log analyst may visualise a reservoir as a multilayered formation, with a given dip, a water/oil contact at the bottom, and a gas/oil contact at the top. But this detailed description of the reservoir by the log analyst does not tell the well test interpreter how the system is going to behave during a well test. He will only know after a test has actually been performed, and the corresponding well test interpretation model will have to be identified from rate and pressure data alone.

Rate and pressure are the only information we need to find the well test interpretation model, but we do need both. If there are no pressure data, the test obviously cannot be analysed. The same conclusion holds if there are no or insufficient rate data. The other models (from geology and logs) will be used only to decide between alternative well test interpretation models.

The reservoir model on the other hand, must integrate all the

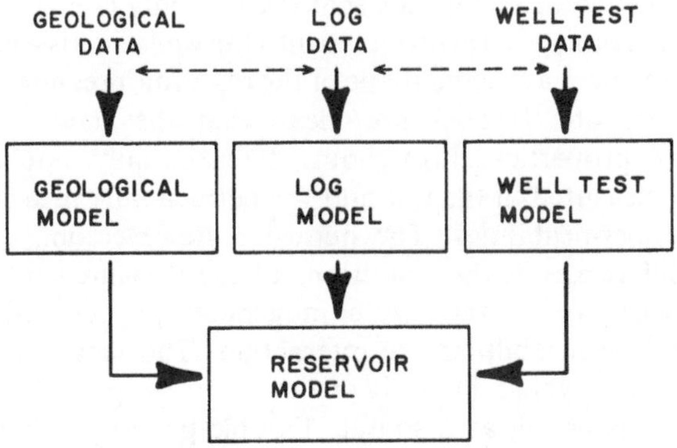

FIG. 3. The concept of interpretation model.

interpretation models obtained from the various sets of data that are available, i.e. the geological model, the log model, and the well test model, not just partial information from these data, as it has been mostly done to date (for instance, geometry from logs, permeability from cores or from tests, etc.). This is the role of the reservoir engineer, and his real challenge. Compared to it, the actual computation of the reservoir model (by finite element, finite differences, etc.) is almost trivial.

2. WELL TEST INTERPRETATION MODELS

2.1. Basic Test Behaviours

Although reservoirs are all different to some extent, the study of well test data in many different reservoirs around the world indicates that the number of possible behaviours during a well test is limited, and only a limited number of interpretation models are required for well test analysis. This results from the fact, already mentioned, that well test interpretation models describe only the behaviour of the reservoir in well testing, and are not a physical description of the reservoir itself.

It appears that all reservoir behaviours can be classified in three main types:

—Homogeneous
—Double porosity
—Double permeability

These basic interpretation models are sketched in Fig. 4.

Homogeneous behaviour means that there is only one porous medium, and that this porous medium produces into the well. This is a mathematical definition: it refers to a specific shape of the reservoir pressure response to a step change in rate. It does not mean that the actual reservoir has homogeneous properties throughout. It essentially implies that the permeability measured in the test and the permeability measured in cores are the same permeabilities. The numbers, however, may be different, because of differences in the conditions of the measurements.

Double porosity means that two homogeneous porous media of distinct porosities and permeabilities are interacting. They may be uniformally distributed, or segregated, but only one medium can produce fluid into the well; the other is acting as a source. Double-porosity behaviour occurs generally in fissured reservoirs, in multilayered reservoirs with high

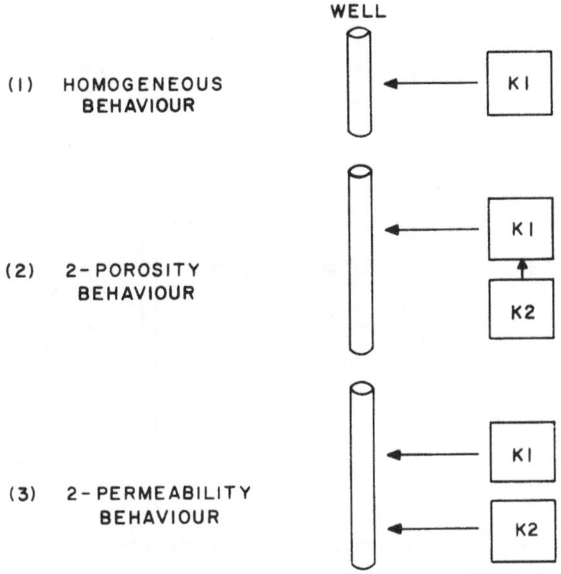

FIG. 4. Basic reservoir behaviours.

permeability contrast between the layers, and in single-layered reservoirs with high permeability variation along the reservoir thickness.

Double permeability refers to two distinct porous media as in double porosity, but each medium can produce into the well. An example of double-permeability behaviour can be found in multilayered reservoirs with low permeability contrast between the layers.

Double porosity and double permeability appear to be the main possible behaviours in heterogeneous systems. Contrary to the homogeneous behaviour, they imply that the permeability measured in a test and the permeability measured in a core may be two different permeabilities. Both should be taken into account in the reservoir model.

2.2. Boundary Effects

These basic test behaviours are rarely seen alone, except sometimes in interference tests, when one measures the pressure response of observation wells away from the active wells. In the most common tests in the oil industry, where measurements are made at the active well, the basic test behaviours are associated with effects of inner and outer boundaries (Fig. 5).

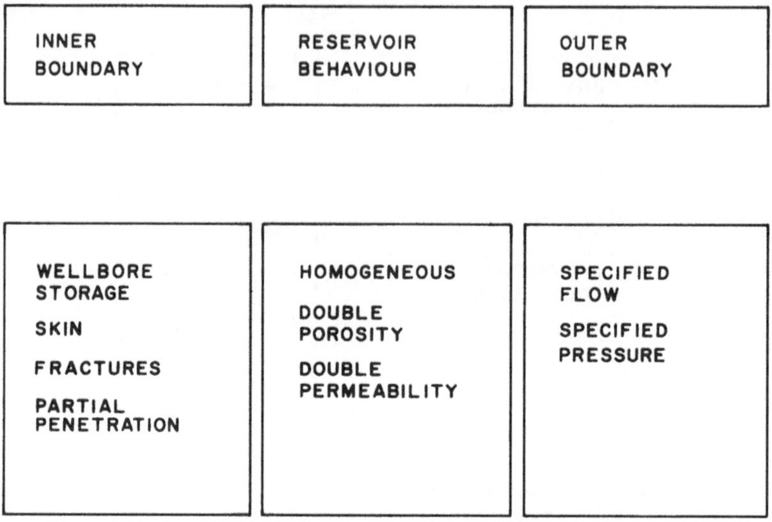

FIG. 5. Components of the interpretation model.

2.2.1. Inner Boundaries

There are only four possible inner boundaries in a reservoir:

—Wellbore storage
—Skin
—Fracture
—Partial penetration

Wellbore storage (also called after flow, anulus unloading, or post-production) characterises the capacity of the wellbore to store or unload fluid. This effect exists in wells with a liquid level, and is easy to visualise in wells filled with liquid: in such wells, during drawdown, the surface flow rate, when the well is first opened for production, is mainly due to the expansion of the fluid stored in the well string and is therefore larger than the sandface flow rate. As time increases, the effect of wellbore storage decreases and the two rates become equal. Similarly, in buildups, the sandface flow rate does not drop to zero instantaneously when the well is shut-in at the surface, because of the compressibility of the fluid in the well column.

Wellbore storage is characterised by a wellbore storage constant C, defined as:

$$C = \mathrm{d}V/\mathrm{d}p \qquad \text{(in bbl/psi)} \tag{1}$$

where dV is the change in the volume of fluid in the wellbore at wellbore conditions (in bbl) and dp is the change in bottomhole pressure (in psi).

For a well filled with a single phase fluid, C can be written as

$$C = c_w V_w \tag{2}$$

where c_w is the compressibility of the fluid in the well string (in psi^{-1}), and V_w is the total wellbore volume (in bbl).

For a well with a changing liquid level, the wellbore storage constant is given by

$$C = \frac{V_u}{\frac{\rho}{144}\frac{g}{g_c}} \tag{3}$$

where V_u is the wellbore volume per unit length (in bbl/ft), ρ is the density of fluid (in lb/ft^3), g is the gravity of the fluid (in ft/s^2) and g_c is a unit conversion factor (in oil field units, $g/g_c = 1{\cdot}0$). Usual values of the wellbore storage constant C are: 0·0001 bbl/psi in tests with downhole shut-in: 0·01 bbl/psi in tests with surface shut-in; and 0·05 bbl/psi in wells with a liquid level. In fissured reservoirs, however, actual wellbore storage values computed from well test data can exceed those computed from eqns (2) and (3) by a factor 10 to 100, because most of the storage is due to the fissure volume.

The *Skin* characterises the condition of the well, in terms of damage resulting from drilling (mud cake) or in terms of improvement resulting from acidising. These wellbore conditions create a steady-state pressure drop in the reservoir (the skin effect) which occurs in an infinitesimal thin zone (the skin) at the sandface.

A zero infinitesimal skin indicates that the zone around the wellbore has the same characteristics as the formation; that is, it is neither damaged nor stimulated.

Positive infinitesimal skin values indicate a damaged well.

A negative infinitesimal skin is an indication of a stimulated well.

Wells with positive infinitesimal skins can be improved through an acid job. Values for infinitesimal skin may range from −4 for an acidised well to +20 for a badly damaged well.

A *Fracture* is defined as a single crack initiated from the wellbore by hydraulic fracturing. Here, we distinguish fractures from fissures, which are natural fractures, and affect the basic reservoir behaviour, not the inner boundary. As we shall see later, pressure behaviour induced by fissures intersecting the well is quite different from that created by a single fracture.

Fractures are characterised by a fracture half-length, x_f, a fracture height (which is often assumed equal to that of the formation interval for vertical fractures), and a fracture conductivity. Hydraulically induced fractures are usually vertical, but may be horizontal if the formation is less than approximately 3000 ft deep.

Fractures created by conventional hydraulic fracturing usually have a very high conductivity, which for all practical purposes, can be considered as infinite. On the other hand, fractures created as a result of massive hydraulic fracturing (MHF), have a finite conductivity. The purpose of MHF is to create very long fractures, with lengths of the order of 1 km. The fracture needs a large quantity of propping agent to be maintained open and as a result, permeability is reduced compared to that of a conventional hydraulic fracture.

Partial penetration occurs when the well has been drilled partially through the producing interval or when only part of a cased well has open perforations. These conditions are characterised by the ratio of the length of the well open to flow to the total length of the well.

Pseudo-skin: Fractures, partial penetration, and fissures intersecting the wellbore create an additional pressure drop not only at the sandface, but also within some distance into the formation. Contrary to that created by an infinitesimal skin, this pressure drop is transient, and becomes steady only after some time of production. This effect is identified as a pseudo-skin. The pseudo-skin associated with partial penetration is always positive (values much greater than 20 are not uncommon) whereas that associated with fractures or fissures is always negative (around -3).

Except with induced fractures, pseudo-skin and infinitesimal skin may coexist. In such a case, the pseudo-skin is always additive to the infinitesimal skin: as a result, a non-damaged well in a fissured reservoir shows a total skin (pseudo plus infinitesimal) of around -3; zero total skin in a fissured reservoir indicates a damaged well whereas the skin for an acidised well could be as negative as -6 or -7. On the other hand, a partially penetrating well would usually show a positive total skin, even after acidification.

Effective wellbore radius: The concept of skin (infinitesimal skin or pseudo-skin) is associated with that of effective wellbore radius:

$$r_{we} = r_w \exp(-S) \tag{4}$$

where S is the total skin. r_{we} is smaller, or larger than r_w if S is positive, zero, or negative, respectively. The effective wellbore radius concept is extremely useful to handle cases where the skin is coupled with other flow regimes.

2.2.2. Outer Boundaries

Test behaviour is also affected by the existence of outer boundaries. These can be of two types:

—specified flow
—specified pressure

3. MODEL IDENTIFICATION

The various components of the interpretation model (inner boundaries, basic behaviour and outer boundaries) are independent from one another, and dominate at different times during the test. They also exhibit distinctly different characteristics. As a result, it is possible to identify each one separately and then construct the complete interpretation model.

This is best accomplished by log–log analysis. The change in pressure during the test, Δp, is plotted versus elapsed time, Δt, on a log–log graph. The log–log scale stresses characteristic shapes and permits identification of dominant flow regimes. The following illustrates some of these characteristic shapes as obtained during a drawdown test in a stabilised reservoir.

3.1. Inner Boundaries

Identification of inner boundaries is performed on early time data.

3.1.1. Wellbore Storage

When wellbore storage dominates, the pressure change, Δp, is proportional to the elapsed time, Δt.[3] As a result, a straight line of unit slope (i.e. one log cycle Δt, for one log cycle Δp) is obtained on a log–log plot (Fig. 6).

A straight line is also obtained when the same points are plotted in cartesian coordinates. This straight line passes through the origin (Fig. 7).

Conversely, a log–log straight line of unit slope through early time data is indicative of wellbore storage only if the same points are also located on a cartesian straight line passing through the origin. A straight line through the same points not passing through the origin might indicate inaccuracy in the evaluation of the initial time or pressure—a common error even with modern electronic gauges.

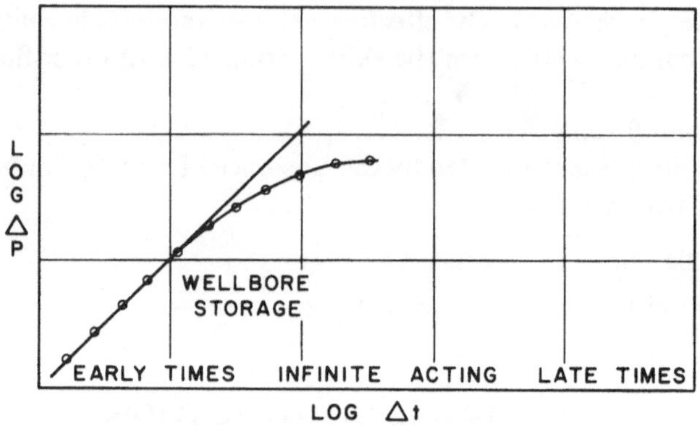

FIG. 6. Wellbore storage log–log behaviour.

3.1.2. High Conductivity Fracture

A high conductivity fracture communicating with the wellbore exhibits an early time log–log straight line of half-unit slope, as Δp is proportional to the square root of Δt during unidimensional flow from the matrix into the fracture (Fig. 8).[4] A straight line passing through the origin is also obtained when data are plotted as Δp versus $\sqrt{\Delta t}$ (Fig. 9).

3.1.3. Low Conductivity Fracture

A low conductivity fracture communicating with the wellbore yields a quarter-unit slope on a log–log plot (Fig. 10) and a straight line passing through the origin on a Δp versus $\sqrt[4]{\Delta t}$ plot (Fig. 11).[5]

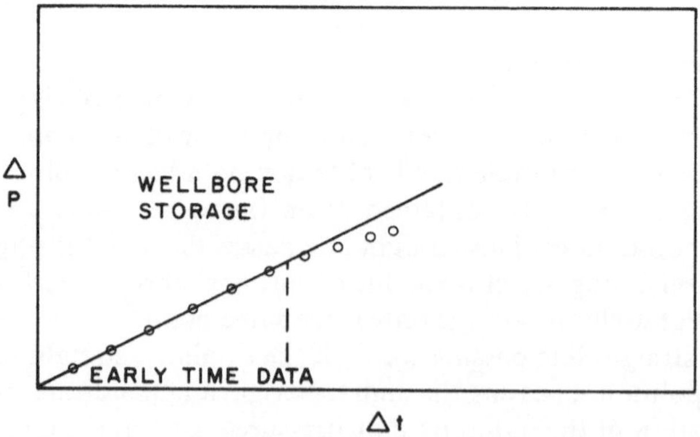

FIG. 7. Wellbore storage: specialised plot.

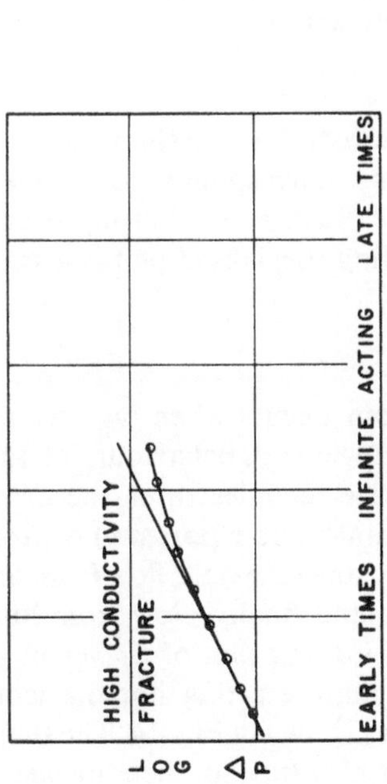

FIG. 8. High conductivity fracture log–log behaviour.

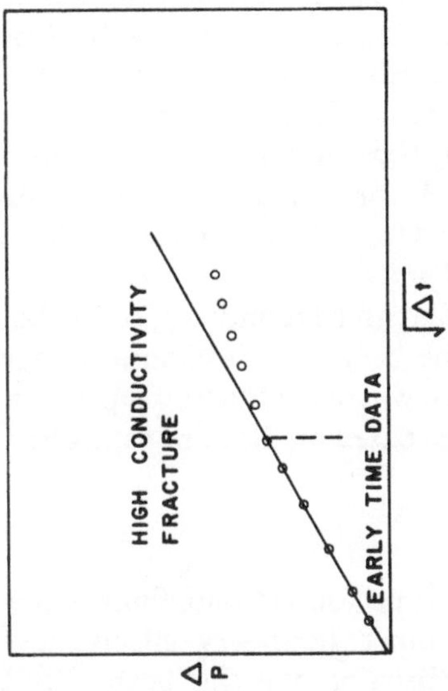

FIG. 9. High conductivity fracture: specialised plot.

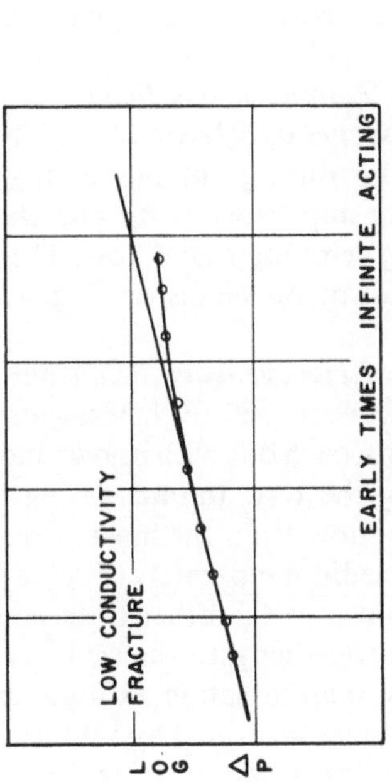

FIG. 10. Low conductivity fracture log–log behaviour.

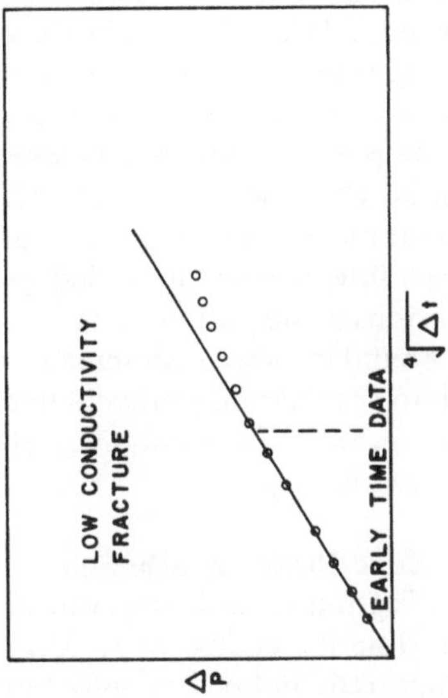

FIG. 11. Low conductivity fracture: specialised plot.

3.1.4. Combination of Inner Boundaries

Wellbore storage, and a high or a low conductivity fracture are examples of inner boundaries that can be distinguished from early time data, because they are revealed by different characteristic behaviours on a log–log plot. However, two different graphs are needed to confirm the diagnostic. The log–log plot is considered a *diagnostic plot* because it is valid for all flow regimes. The other plots, such as Δp versus Δt for wellbore storage, and Δp versus the square or the fourth root of Δt for a fracture, which are specific to a given flow regime, are called *specialised plots*.

Various combinations of inner boundaries can be found in practice, but one boundary always dominates early time behaviour: wellbore storage tends to dominate any other boundary effect, whereas a fracture supersedes partial penetration. Analysis must always be based on the dominating flow regime (Fig. 13).

3.2. Basic Reservoir Behaviour

Identification of basic reservoir behaviour is performed on infinite acting data. Infinite acting flow occurs after inner boundary effects have disappeared, and before outer boundary effects become significant. Basic reservoir behaviour dominates during infinite acting.

3.2.1. Homogeneous Behaviour

Homogeneous behaviour exhibits the characteristic log–log shape shown in Fig. 12 during infinite acting. This shape corresponds to a linear relationship between Δp and the log of Δt.[6] Such a relationship occurs during semi-log radial flow. The corresponding specialised plot is a semi-log plot of Δp versus Δt (Fig. 13).

3.2.2. Heterogeneous Behaviour

As discussed before, heterogeneous behaviour occurs when two distinct systems, each one with its own individual homogeneous behaviour, interact during the test. Infinite acting in such a case includes three successive stages: flow from the most permeable medium (double porosity) or from each medium separately (double permeability) into the well; flow from one medium into the other (interporosity flow); and finally, flow from both media together into the well.[7] As a result, a log–log plot of Δp versus Δt during infinite acting exhibits two distinct infinite acting homogeneous behaviours separated by a transition period, as shown in Fig. 14. The shape of the transition provides information on the type of heterogeneous behaviour.

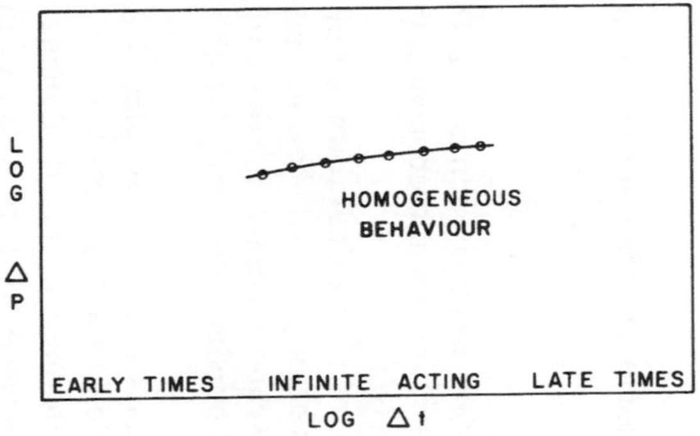

FIG. 12. Infinite acting homogeneous log–log behaviour.

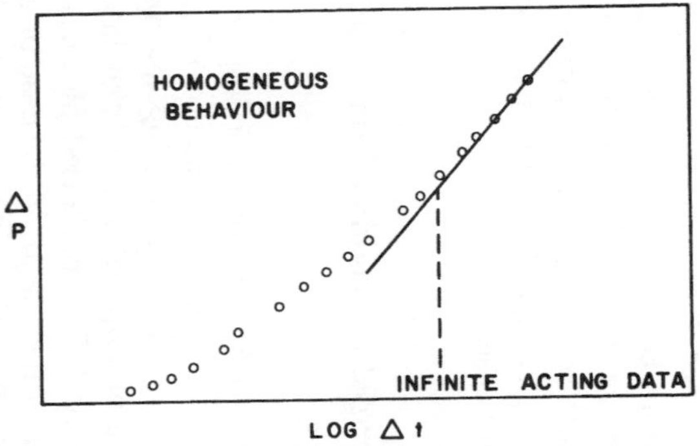

FIG. 13. Infinite acting homogeneous behaviour: specialised plot.

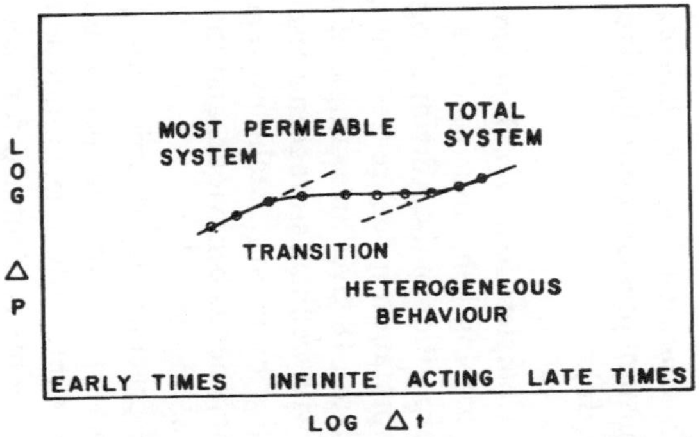

FIG. 14. Infinite acting heterogeneous log–log behaviour.

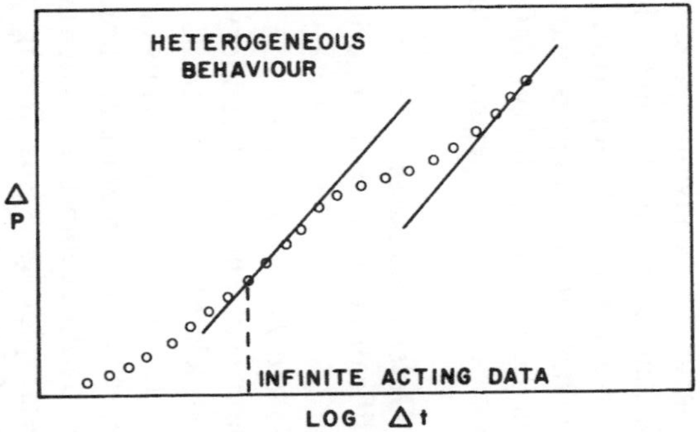

FIG. 15. Infinite acting heterogeneous behaviour: specialised plot.

The specialised plot is a semi-log plot of Δp versus Δt, as in the homogeneous behaviour case. However, since there are two distinct homogeneous behaviours in the heterogeneous behaviour case, there may exist two distinct semi-log straight lines on the specialised plot (Fig. 15). These straight lines may be parallel (double porosity) or have different slopes (double permeability). A third semi-log straight line may also exist during the transition period in some special cases.[8]

3.3. Outer Boundaries

Identification of outer boundaries is performed on late time data.

3.3.1. No-Flow Boundary

In the case of a no-flow boundary, the Δp versus Δt curve on a log–log plot exhibits a late time increase in slope with respect to infinite acting flow (Fig. 16). In the case of a completely closed system, for instance, the log–log Δp versus Δt curve becomes asymptotic to a unit slope straight line at late times, as Δp becomes a linear function of Δt.[9] The specialised plot is a cartesian plot Δp versus Δt (Fig. 17). On the other hand, in the case of a sealing fault, Δp is proportional to $\log \Delta t$ at late times, and the specialised plot is a semi-log plot of Δp versus Δt, which exhibits a late time semi-log straight line with a slope equal to twice that for infinite acting flow.[10]

3.3.2. Constant Pressure Boundary

A constant pressure boundary is characterised by a stabilisation in pressure (Fig. 16).

3.4. Complete Reservoir Behaviour

The log–log behaviour of the actual reservoir is simply obtained as the superposition of the individual log–log behaviour of each component of the interpretation model representing the reservoir.

Figure 18 gives an example of a log–log trace for a drawdown test of a well with wellbore storage in a reservoir with homogeneous behaviour and a no-flow boundary. Each flow regime (i.e. wellbore storage, infinite acting homogeneous behaviour, and the no-flow boundary) are clearly identified on the log–log plot.

In practice, the identification process starts with a log–log plot of Δp versus Δt for the flow period under consideration. The log–log plot is divided into early time, infinite acting, and late time data, and the dominant flow regime is identified for each of these groups from the shape of the log–log trace. This defines the interpretation model for the test. The

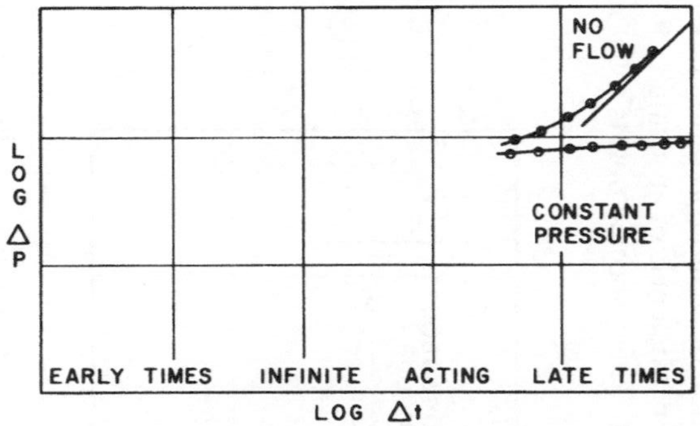

FIG. 16. Outer boundary log–log behaviour.

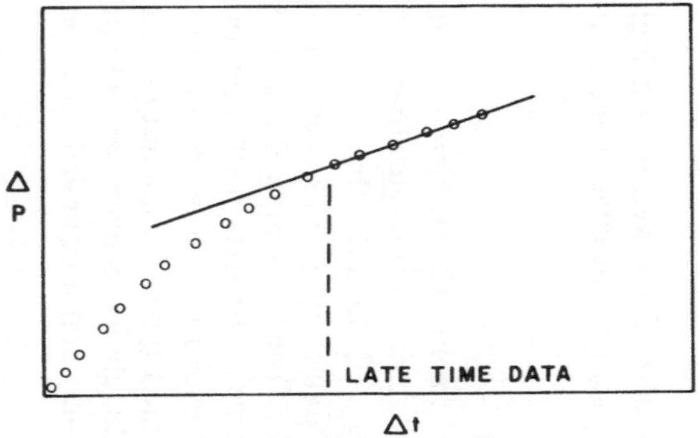

FIG. 17. Closed outer boundary: specialised plot.

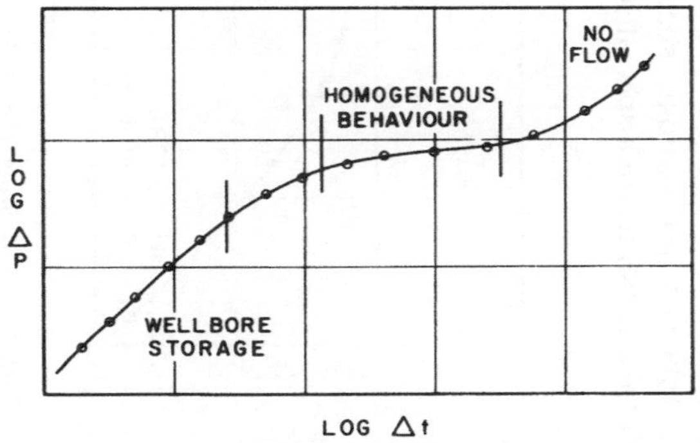

FIG. 18. Log–log behaviour of a well with wellbore storage in a closed reservoir with homogeneous behaviour.

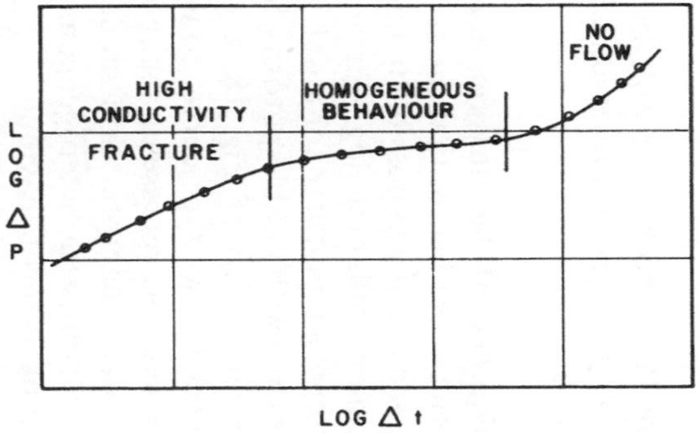

FIG. 19. Log–log behaviour of a well with a high conductivity fracture in a closed reservoir with homogeneous behaviour.

pressure–time relationship for the interpretation model is called a 'type-curve'.

Figure 19 is another example of a type-curve, for a well with a high conductivity fracture in a reservoir with homogeneous behaviour and a no-flow boundary.

3.5. Uniqueness of the Interpretation Model

It has been noted before that the solution of the inverse problem was not unique, and that it was possible to find several reservoir configurations that could exhibit similar log–log behaviour. Figure 20 illustrates this difficulty. It represents a type-curve for a well with wellbore storage in a reservoir of infinite extent with double-porosity behaviour. The type-curve of Fig. 20 is very similar to that of Fig. 18, and for some combination of reservoir parameters, the two type-curves could become identical (upper part of Fig. 21). This example represents the most common type of ambiguity in the selection of an interpretation model, i.e. the choice between an infinite acting reservoir with heterogeneous behaviour, and a bounded reservoir with homogeneous behaviour.

3.5.1. Consistency

It is usually possible to eliminate an incorrect interpretation model by evaluating its overall consistency. The selection of an interpretation model has a number of implications, such as the existence of a straight line through specific points on a specialised plot. These specific points must be

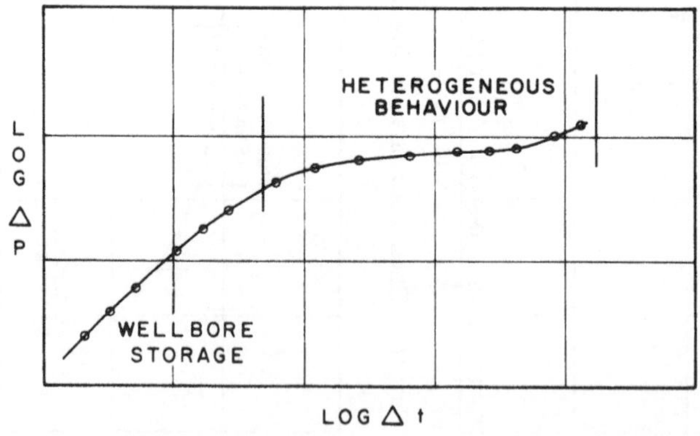

FIG. 20. Log–log behaviour of a well with wellbore storage in an infinite reservoir with double-porosity behaviour.

exactly the same as indicated on the log–log plot for the particular flow regime under consideration. If they are different, the selected model is inconsistent and, therefore, incorrect.

3.5.2. Differential of Pressure Curve

One efficient way to distinguish between homogeneous and heterogeneous basic behaviour is to examine on a log–log plot, the derivative of Δp with respect to the natural log of Δt, in addition to the Δp versus Δt curve. The shape of the derivative is drastically different in the two cases[11,12] as illustrated in the lower part of Fig. 21. Figure 21 presents log–log traces for a well with wellbore storage in: (1) an infinite reservoir with homogeneous behaviour; (2) a reservoir with homogeneous behaviour bounded by a sealing fault; and (3) an infinite reservoir with double-porosity behaviour.

The log–log traces of Δp versus Δt are superposed in the upper part of Fig. 21. They are very similar, and it is very difficult to distinguish between the three interpretation models, especially if the test is not long enough for the features at the extreme right of the plot to be clearly seen. On the other hand, the difference between the three interpretation models becomes

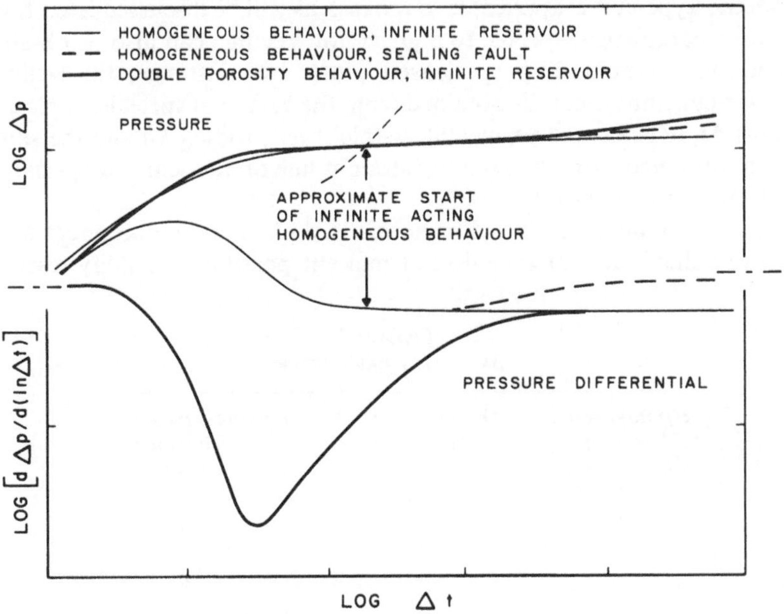

FIG. 21. Log–log behaviour of differential of pressure curve.

obvious on the log–log traces of the differential of the pressure curves, $d\Delta p/d(\ln \Delta t)$ versus Δt, shown in the lower part of Fig. 21. Heterogeneous behaviour exhibits a characteristic hump during transition which allows unambiguous identification, provided the quality of the pressure data is adequate.

4. ESTIMATION OF MODEL PARAMETERS

4.1. Basic Test Parameters

Once the well test interpretation model has been identified, it is possible to compute its characteristic parameters, which are assumed to represent the corresponding well and reservoir parameters. The number of parameters that can be extracted from a test is determined by the components of the interpretation model, but the exact meaning of these parameters is controlled by the well behaviour (homogeneous versus heterogeneous).

As an example, Table 1 lists all the parameters that can be obtained in the case of a well with wellbore storage in a reservoir of infinite extent, for the various possible reservoir behaviours. All these parameters can be obtained from the 'type-curve' applicable to the test, as will be discussed later. Some of these parameters, specific to a given flow regime, can also be obtained from the corresponding specialised plots. For instance, the wellbore storage constant C can be obtained from the cartesian specialised plot, Δp versus Δt, whereas the permeability–thickness product kh and the skin S can be obtained from the semi-log straight line on the semi-log specialised plot, Δp versus $\log \Delta t$.

Specialised analyses, based on specialised plots, are independent of the overall behaviour, but they do not make it possible to qualify reservoir

TABLE 1
BASIC TEST PARAMETERS

Homogeneous behaviour	*Double-porosity behaviour*	*Double-permeability behaviour*
kh	(kh) 1	(kh) t
S	S	S
C	C	C
	λ	λ
	ω	ω
		κ

parameters obtained from specialised analyses alone. This represents the main difference between the methodology presented here, and the so-called 'conventional' analysis techniques,[13] which are limited to the use of specialised plots, and in particular, of the semi-log plot, without the identification of the complete interpretation model. As a result, the interpretation process with these techniques is incomplete at best, and can yield misleading results at worst.

The ambiguity of relying only on specialised analyses is illustrated in Table 1. For instance, in the case of homogeneous behaviour, kh simply represents the permeability–thickness product of the reservoir. On the other hand, in the case of double-porosity behaviour, kh is the permeability–thickness product for the most permeable system only, whereas it represents the total system in the case of double-permeability behaviour.

The importance of identifying the reservoir behaviour is even more evident for the skin S. As mentioned earlier, in the case of homogeneous behaviour, the skin is positive for a damaged well, zero for a non-damaged well and negative for a stimulated well. In the case of double-porosity behaviour, however, a normal well (not damaged or stimulated) exhibits a skin of -3 (in reality, a pseudo-skin). Therefore, a zero skin indicates a damaged well, while a skin of -3 indicates that the well can be further stimulated, to a skin of -7. Appropriate action on the well cannot, therefore, be based on the value of the computed skin alone, without the knowledge of the complete interpretation model and reservoir behaviour.

Identification of the interpretation model also guarantees that all available information is extracted from the test data. As indicated in Table 1, for the same inner and outer boundary conditions, a reservoir with heterogeneous behaviour requires more parameters for characterisation than one with a simple homogeneous behaviour (namely, the storativity ratio ω, the interporosity flow coefficient λ, and the permeability ratio κ).

4.2. Type-Curve Analysis

After all the components of the interpretation model have been identified from log–log analysis, it is possible to construct a theoretical model that includes these components and represents the behaviour of the reservoir under study (solution of the inverse problem). Reservoir parameters are then calculated by adjusting the parameters of this specific theoretical model until its response matches the actual test measurements (solution of the direct problem).

In practice, the response of the theoretical model is presented as a family of *dimensionless type-curves*, expressed in terms of a *dimensionless pressure* and a *dimensionless time*, with each curve in the family being characterised by dimensionless coefficients. These dimensionless parameters are defined as the product of a measured parameter (e.g. Δp or Δt) and parameters characterising the reservoir (porosity ϕ, permeability–thickness kh, well radius $r_w, \ldots$), the fluid (viscosity μ, formation volume factor B, total compressibility $c_t, \ldots$) and the test (flow rate $q, \ldots$). For instance, the dimensionless pressure p_D is expressed as

$$p_D = A\Delta p \qquad \text{where } A = f_p(kh, q, B, \phi, \ldots) \tag{5}$$

while the dimensionless time t_D is expressed as

$$t_D = B\Delta t \qquad \text{where } B = f_t(k, \phi, \mu, c_t, r_w, \ldots) \tag{6}$$

In log–log coordinates, these become:

$$\log p_D = \log \Delta p + \log A$$
$$\log t_D = \log \Delta t + \log B$$

which points out that, *when the appropriate interpretation model has been selected*, the actual pressure versus time curve, Δp versus Δt, and the theoretical one, p_D versus t_D, have identical shapes but are shifted with respect to one another when plotted on the same log–log graph.

The objective of *type-curve analysis*[14] is to evaluate the amount of shift for both pressure and time axes. These are then used to compute the corresponding reservoir parameters. The most common procedure is to plot actual data, Δp versus Δt, on a log–log graph, and superimpose this graph on the appropriate theoretical type-curve drawn on a log–log graph with identical cycle lengths. The graph with the actual curve is shifted over that with the theoretical type-curve, while keeping the time and pressure axes parallel on each graph, until a good match is found between the actual data and a theoretical curve. When this is achieved, a match point is selected, and its coordinates recorded on each graph. The graph with the theoretical type-curve provides the values of p_D and t_D for the match point, whereas the graph with the actual data provides the values of Δp and Δt. Reservoir parameters are obtained by substituting p_D and Δp and t_D and Δt, into eqns (5) and (6), respectively. Additional reservoir information is obtained from the value of the dimensionless parameter corresponding to the matched theoretical curve.

For a given interpretation model, however, not all type-curves are equivalent. A non-ambiguous match can only be obtained if the dimensionless parameters represent independent groups of variables.

Furthermore, depending upon the choice of the dimensionless parameters, one type-curve may be easier to use within a specific time range. Dimensionless parameters must be selected so that all the various theoretical curves on the type-curve graph merge into one single curve where the actual well data are the most numerous. In addition, the range of the various flow regimes must be clearly indicated, with limits computed from realistic approximation criteria, so that the appropriate *specialised analyses* can be applied to the corresponding data for checking the consistency of the analysis.[1]

4.2.1. Drawdown Type-Curves

Type-curves used for the analysis of an initial production test in a reservoir previously at a uniform pressure, p_i, are called *drawdown type-curves*. They are defined as

$$p_D = p_D[(\Delta t)_D] \tag{7}$$

Actual data for type-curve analysis are defined as

$$\Delta p = p_i - p(\Delta t) \tag{8}$$

versus Δt, where Δt is the elapsed time since the start of the test sequence, and $p(\Delta t)$ the corresponding pressure reading (Fig. 22).

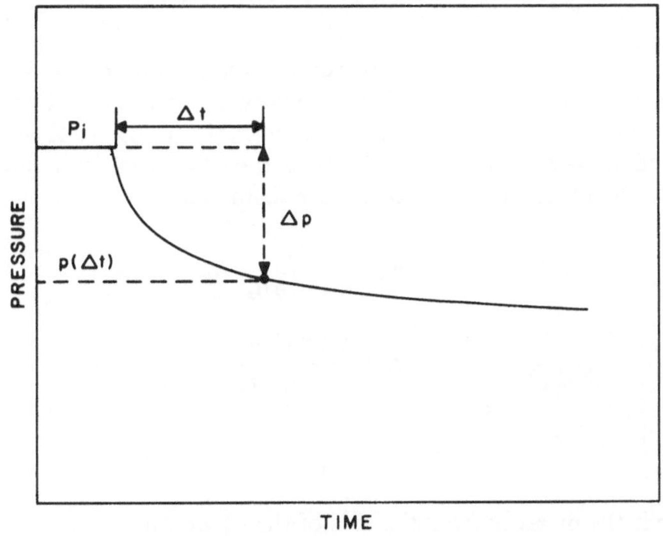

FIG. 22. Drawdown pressure test.

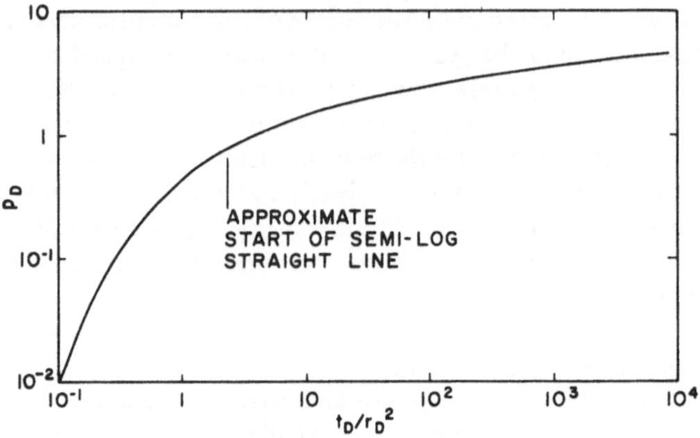

FIG. 23. Type-curve for an observation well in an infinite reservoir with homogeneous behaviour.

4.2.2. Published Drawdown Type-Curves

All possible interpretation models can be obtained by a simple combination of the various components (four inner boundaries, three basic behaviours, and two outer boundaries), which amounts to only a few score of theoretical type-curves. Some have been published in the literature for the most common reservoir configurations. They are described hereafter.

4.2.2.1. Type-curve for an observation well in an infinite reservoir with homogeneous behaviour (Fig. 23).[14] This type-curve is based on the well-known line-source solution, and is applicable to the analysis of data measured in observation wells away from the production well (interference testing). Dimensionless parameters are defined as:

$$p_D = \frac{kh}{141 \cdot 2qB\mu} \Delta p \tag{9}$$

$$t_D = \frac{0 \cdot 000\,264k\,\Delta t}{\phi\mu c_t r_w^2} \tag{10}$$

and

$$r_D = r/r_w \tag{11}$$

where r is the distance from the axis of the observation well to that of the production well. All parameters are expressed in engineering oil field units.

As indicated from Fig. 23, a single curve is obtained when using the independent groups:

$$p_D \qquad t_D/r_D^2$$

Type-curve analysis of Δp versus Δt yields:

1. The permeability thickness product kh from the pressure match (eqn (9)).
2. The reservoir storativity $\phi c_t h$ from the time match (eqn (10)).

The values kh and $\phi c_t h$ represent the total amount of information that can be extracted from interference tests in formation with homogeneous behaviour in the absence of boundary effects.

ASSOCIATED SPECIALISED PLOTS: Infinite acting radial flow starts approximately at $t_D/r_D^2 = 3$. Actual pressure points matching the type-curve beyond $t_D/r_D^2 = 3$ are located on a semi-log straight line on the semi-log specialised plot.

4.2.2.2. Type curve for a well with wellbore storage and skin in an infinite reservoir with homogeneous behaviour (Fig. 24).[1] Dimensionless parameters are defined as

$$p_D \text{ from eqn (9)} \qquad t_D \text{ from eqn (10)}$$

$$C_D = \frac{0{\cdot}8936C}{\phi c_t h r_w^2} \tag{12}$$

A single family of curves is obtained when using the independent groups:

$$p_D$$

$$t_D/C_D = 0{\cdot}000\,295\,\frac{kh}{\mu}\,\frac{\Delta t}{C} \tag{13}$$

with each curve in the family represented by

$$C_D \exp(2S) = \frac{0{\cdot}8936C\exp(2S)}{\phi c_t h r_w^2} \tag{14}$$

Type-curve analysis of Δp versus Δt yields:

1. The permeability–thickness product kh from the pressure match (eqn (9)).
2. The wellbore storage constant C from the time match (eqn (13)).
3. The skin factor S from the curve match (eqn (14)).

The values kh, C and S represent the total amount of information that can be extracted from the test, in the absence of boundary effects. In addition, it is possible to evaluate the quality of the well from the $C_D \exp(2S)$ value. In general, damaged wells correspond to $C_D \exp(2S)$ values greater than 1000, non-damaged wells to $C_D \exp(2S)$ values between 5 and 1000, acidised wells to $C_D \exp(2S)$ values between 0·5 and 5, and fractured wells to $C_D \exp(2S)$ values less than 0·5.

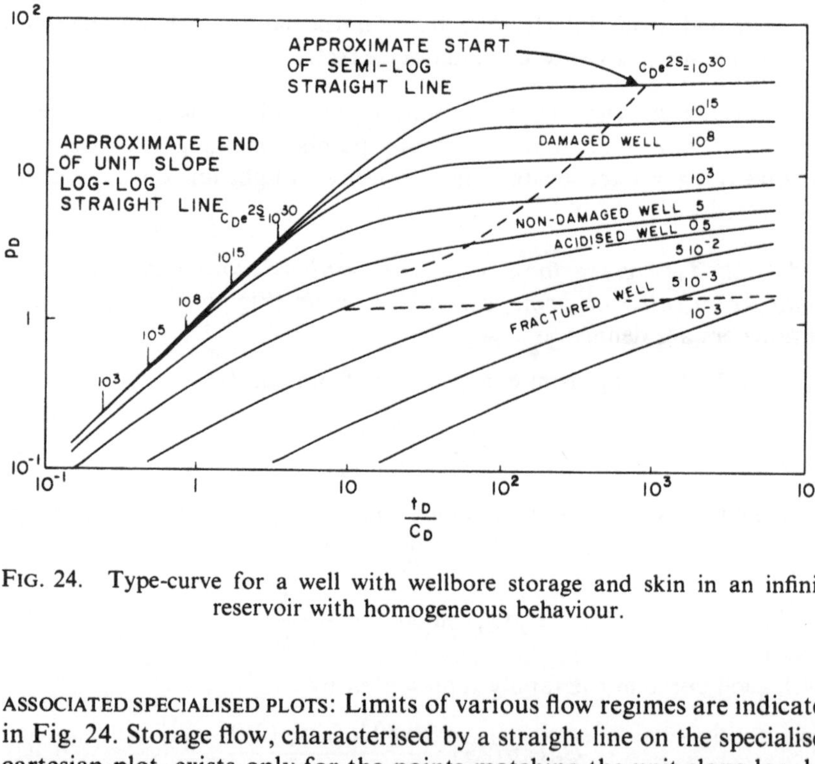

FIG. 24. Type-curve for a well with wellbore storage and skin in an infinite reservoir with homogeneous behaviour.

ASSOCIATED SPECIALISED PLOTS: Limits of various flow regimes are indicated in Fig. 24. Storage flow, characterised by a straight line on the specialised cartesian plot, exists only for the points matching the unit slope log–log straight line on Fig. 24. In practice, this can only occur if the well is damaged ($C_D \exp(2S)$ greater than 10^3). An early time cartesian straight line does not exist if the well is non-damaged or stimulated.

The approximate start of infinite acting radial flow is indicated by a curve in Fig. 24. Pressure points matching the drawdown type curve beyond that curve are located on a semi-log straight line on the semi-log specialised plot.

4.2.2.3. Type curve for a well with an infinite conductivity vertical fracture at the centre of a closed rectangular reservoir with homogeneous behaviour (Fig. 25).[15] Dimensionless parameters are defined as

$$p_D \text{ from eqn (9)}$$

$$t_{Df} = \frac{0{\cdot}000\,264 k \Delta t}{\phi \mu c_t (x_f)^2} \tag{15}$$

$$x_f/\sqrt{A} \tag{16}$$

$$x_e/y_e \tag{17}$$

where x_f represents the fracture half-length, A the area of the rectangular reservoir, x_e the half-length of the rectangular rêservoir, parallel to the fracture, and y_e the half-width (Fig. 26). These parameters are independent and yield a single family of curves.

Type-curve analysis of Δp versus Δt yields:

1. The permeability–thickness product kh from the pressure match (eqn (9)).
2. The fracture half-length x_f from the time match (eqn (15)).
3. The reservoir area A and length-to-width ratio x_e/y_e from the curve match (eqns (16) and (17)).

The values kh, x_f, A and x_e/y_e represent the total amount of information that can be extracted from the test.

ASSOCIATED SPECIALISED PLOTS: Limits of the various flow regimes are indicated in Fig. 25. Fracture flow, characterised by a straight line on the specialised Δp versus Δt plot, exists only at very early times for the points matching the half-unit slope log–log straight line on the $(x_f/\sqrt{A} = 0)$ curve in Fig. 25, which corresponds to a reservoir of infinite extent.

The approximate start of infinite acting radial flow occurs at approximately $t_{Df} = 2$ on the infinite reservoir curve. Pressure points matching the infinite reservoir drawdown type curve beyond $t_{Df} = 2$ are located on a semi-log straight line on the semi-log specialised plot. A cartesian specialised plot yields a straight line for pressure points matching the $(x_f/\sqrt{A})$ curves at late times.

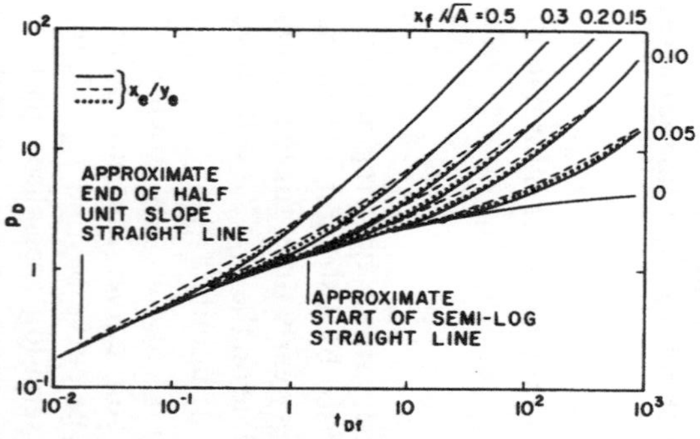

FIG. 25. Type-curve for a well with an infinite conductivity vertical fracture at the centre of a closed rectangular reservoir with homogeneous behaviour.

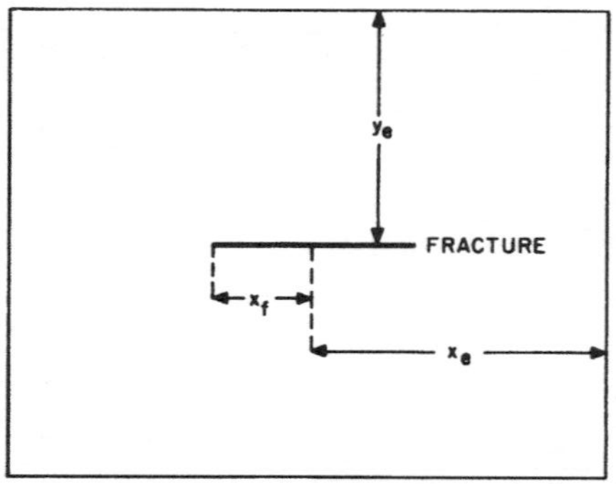

FIG. 26. Schematic of a vertically fractured well in a rectangular reservoir.

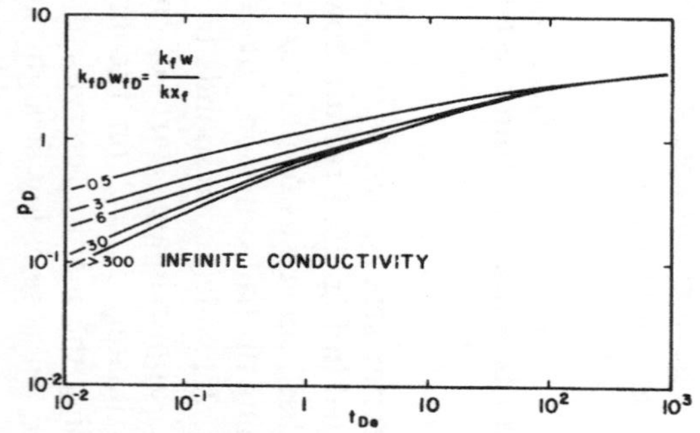

FIG. 27. Type-curve for a well with a finite conductivity vertical fracture in an infinite reservoir with homogeneous behaviour.

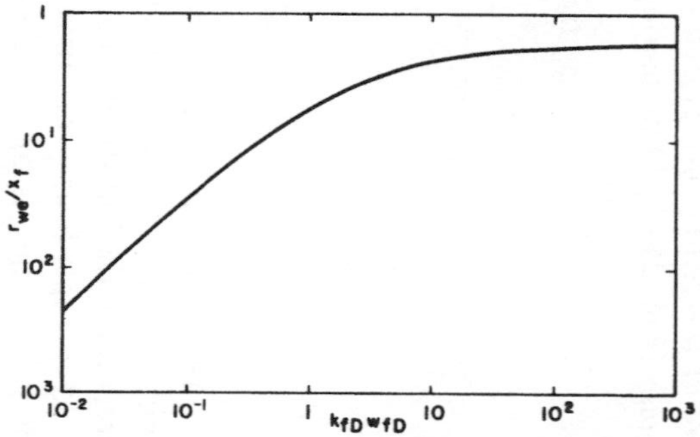

FIG. 28. Relation between equivalent wellbore radius and fracture finite conductivity.

4.2.2.4. Type-curve for a well with a finite conductivity vertical fracture in an infinite reservoir with homogeneous behaviour (Fig. 27).[5] Dimensionless parameters are defined as:

$$p_D \text{ from eqn (9)}$$

$$t_{De} = \frac{0{\cdot}000\,264 k \Delta t}{\phi \mu c_t (r_{we})^2} \tag{18}$$

$$k_{fD} w_{fD} = k_f w / k x_f \tag{19}$$

where r_{we} is the effective well radius, k_f the fracture conductivity, and w the fracture thickness. These parameters are independent and yield a single family of curves. The relationship between r_{we}, the fracture half-length x_f and the dimensionless fracture conductivity $k_{fD}w_{fD}$ is given in Fig. 28 ($r_{we} = x_f/2$ for infinite fracture conductivity).

Type curve analysis of Δp versus Δt yields:

1. The permeability–thickness product kh from the pressure match (eqn (9)).
2. The effective well radius r_{we} from the time match (eqn (18)).
3. The dimensionless fracture conductivity $k_{fD}w_{fD}$ from the curve match.
4. The fracture half-length x_f from Fig. 28.
5. The actual fracture conductivity $k_f w$ from eqn (19).

The values kh, x_f and $k_f w$ represent the total amount of information that can be extracted from the well in the absence of boundary effects.

ASSOCIATED SPECIALISED PLOTS: Limits of the various flow regimes are not indicated in Fig. 27. Finite conductivity fracture flow, characterised by a straight line on a specialised Δp versus $\sqrt[4]{\Delta t}$ plot exists only at early times, on the $k_{fD}w_{fD}$ curves that exhibit a quarter-slope log–log straight line. Infinite acting radial flow characteristics are only seen for pressure points matching the lower curve, which represents infinite conductivity, for t_{De} values greater than 20.

4.2.2.5. Type curves for an observation well in an infinite reservoir with double porosity behaviour.[8] There are two different type-curves, depending upon the nature of interporosity flow. Figure 29 corresponds to restricted interporosity flow, where there is a skin between matrix and fissures, as

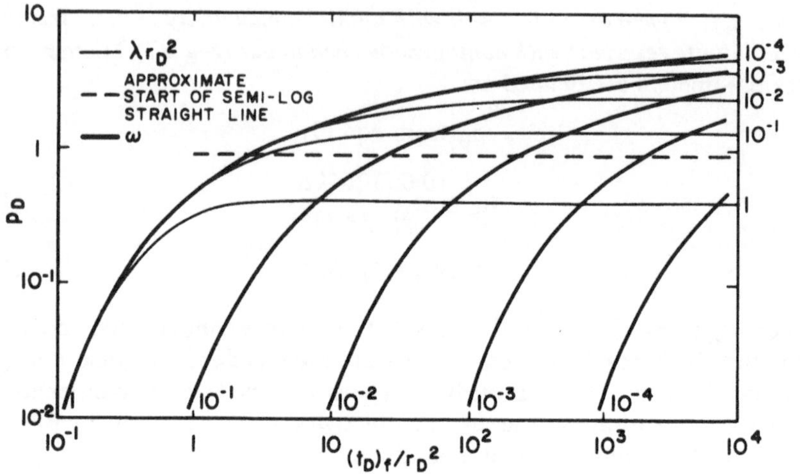

FIG. 29. Type-curve for an observation well in an infinite reservoir with double-porosity behaviour (restricted interporosity flow).

would be created by calcite deposit, for instance. Figure 30 corresponds to unrestricted interporosity flow.

p_D from eqn (9)

$$(t_D)_f = \frac{0{\cdot}000\,264 k_f \Delta t}{(\phi V c_t)_f \mu r_w^2} \tag{20}$$

r_D from eqn (11)

$$\omega = \frac{(\phi V c_t)_f}{(\phi V c_t)_f + (\phi V c_t)_m} = \frac{(\phi V c_t)_f}{(\phi V c_t)} \tag{21}$$

$$\lambda = \alpha r_w^2 \frac{k_m}{k_f} \tag{22}$$

Subscripts f and m refer to the most permeable and the least permeable medium, respectively; f + m refers to the total system; $(\phi V c_t)$ is the storativity, V being the concentration of one medium (i.e. the ratio of the volume of that medium to the bulk volume) and ϕ the porosity of that medium (i.e. the ratio of pore space in the medium to the volume of the medium); ω is the ratio of the storativity of the most permeable medium to the total storativity; λ is the interporosity flow coefficient; α is a geometrical

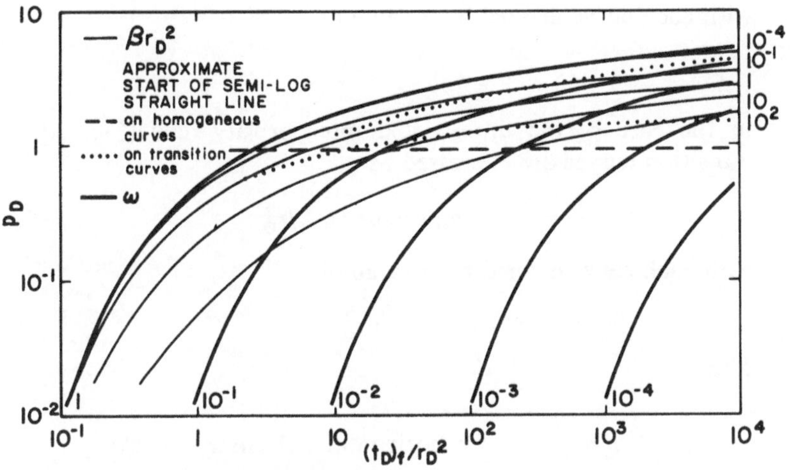

FIG. 30. Type-curve for an observation well in an infinite reservoir with double-porosity behaviour (unrestricted interporosity flow).

factor which depends upon the shape of the matrix blocks constituting the least permeable medium, i.e.

$$\alpha = \frac{12}{h_m^2} \quad \text{(for horizontal slab matrix blocks of thickness } h_m) \quad (23)$$

$$\alpha = \frac{15}{r_m^2} \quad \text{(for sperical matrix blocks of radius } r_m) \quad (24)$$

The type-curves shown in Figs 29 and 30 are obtained as the superposition of two families of curves:

—One family represents line source solutions for reservoirs with homogeneous behaviour, as in Fig. 23, and are expressed as

$$p_D \text{ versus } (t_D)_f/r_D^2$$

each curve being characterised by a value of ω. The value $\omega = 1$ represents the behaviour of the most permeable system, which is the first to appear in the test data. The other ω curves represent the behaviour of the total system, after the end of the transition period.

—The other family represents transition curves:
In the case of restricted interporosity flow (Fig. 29), the transition curves are expressed as

$$p_D \text{ versus } (t_D)_f/r_D^2$$

with each curve defined by a value of

$$\lambda r_D^2$$

In the case of the unrestricted interporosity flow (Fig. 30), the transition curves are expressed as

$$p_D \text{ versus } (t_D)_f / r_D^2$$

with each curve defined by a value of

$$\beta r_D^2$$

where

$$\beta = \frac{\lambda}{3\omega} \quad \text{for horizontal slab matrix blocks} \tag{25}$$

$$\beta = \frac{3\lambda}{5\omega} \quad \text{for spherical matrix blocks} \tag{26}$$

The two families of transition curves exhibit sufficiently different shapes on a log–log plot so that there is usually little ambiguity in differentiating between them. However, they represent two extremes. There is in theory a wide range of possible transitions, depending upon the magnitude of the skin between the most permeable and the least permeable media. In practice, the restricted interporosity flow behaviour appears to be most common.

Type-curve matching involves three curves, sequentially:

—first, the $\omega = 1$ homogeneous curve for pressure data corresponding to the response of the most permeable system;
—second, one of the λr_D^2 or βr_D^2 curves for pressure data corresponding to the transition period;
—third, one of the ω homogeneous curves for pressure data corresponding to the response of the total system.

Because the type-curves in Figs 29 and 30 do not represent exact double porosity behaviour, but only its various components, the match during transition cannot be perfect, especially for high ω values. The transition curve to be used for quantitative analysis is the one passing through the mid-point of the transition zone.

Type-curve analysis of Δp versus Δt yields

1. The permeability thickness product kh from the pressure match (eqn (9)); kh represents, in fact, $k_f h$, the permeability–thickness for the most permeable system (the least permeable system does not produce into the well).
2. The storativity of the most permeable system $(\phi V c_t)_f h$, from the time match (eqn (20)).
3. ω from the total system curve match.
4. The storativity of the least permeable system, $(\phi V c_t)_m h$, from eqn (21)).
5. λ from the transition curve match (λr_D^2 or βr_D^2, depending upon the nature of interporosity flow). In the case of unrestricted interporosity flow, the shape of the matrix blocks must be known or must be assumed.
6. The permeability of the least permeable system, k_m, from eqn (22).

$k_f h$, $(V c_t)_f h$, and λ represent the total amount of information that can be extracted from interference tests in formations with double-porosity behaviour in the absence of boundary effects.

ASSOCIATED SPECIALISED PLOT: The start of infinite acting radial flow on each homogeneous curve is indicated by a horizontal continuous line corresponding approximately to $p_D = 1$ in Figs 29 and 30. Actual pressure points matching a ω homogeneous curve beyond this line are located on a semi-log straight line on the semi-log specialised plot. If this occurs for both initial and final homogeneous curves, two semi-log straight lines are present on the specialised plot. These two straight lines are parallel because they both relate to the permeability of the most permeable system. The occurrence of the first semi-log straight line is rare in practice, however. In the case of unrestricted interporosity flow (Fig. 30), a third semi-log straight line may exist if actual pressure points during transition match a βr_D^2 transition curve inside the region limited by two dotted curves in Fig. 30. This transition semi-log straight line has a slope equal to half that of the semi-log straight lines corresponding to infinite acting radial flow.

4.2.2.6. Type-curves for a well with wellbore storage and skin in an infinite reservoir with double-porosity behaviour.[8] As for interference testing, there are two different type-curves, depending upon the nature of interporosity flow. Figure 31 corresponds to restricted interporosity flow, while Fig. 32 corresponds to unrestricted interporosity flow.

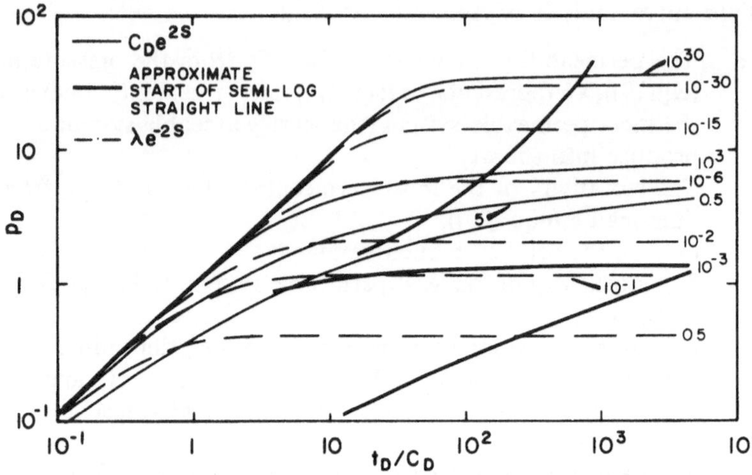

FIG. 31. Type-curve for a well with wellbore storage and skin in a reservoir with double-porosity behaviour (restricted interporosity flow).

Dimensionless parameters are defined as

p_D from eqn (9)

$$(t_D)_{f+m} = \frac{0{\cdot}000\,264 k_f \Delta t}{(\phi V c_t)_{f+m} \mu r_w^2} \tag{27}$$

$$C_D = \frac{0{\cdot}8936C}{(\phi V c_t)_{f+m} h r_w^2} \tag{28}$$

ω from eqn (21)

λ from eqn (22)

The type-curves shown in Figs 31 and 32 are obtained as the superposition of two families of curves:

—one family corresponds to the homogeneous behaviour of both most permeable and total systems, and is represented by the type-curves for a well with wellbore storage and skin in an infinite reservoir with homogeneous behaviour of Fig. 24:

$$p_D = f\left(\frac{t_D}{C_D}\, C_D \exp(2S)\right)$$

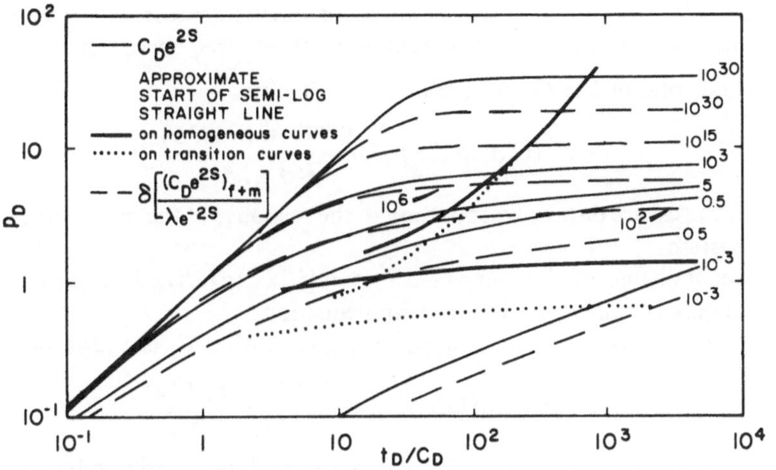

FIG. 32. Type-curve for a well with wellbore storage and skin in a reservoir with double-porosity behaviour (unrestricted interporosity flow).

—the other family represents transition curves:

In the case of restricted interporosity flow (Fig. 31), the transition curves are expressed as

$$p_D \text{ (eqn (9)) versus } t_D/C_D \text{ (eqn (13))}$$

with each curve defined by a value of

$$\lambda \exp(-2S)$$

In the case of unrestricted interporosity flow (Fig. 32), the transition curves are expressed as

$$p_D \text{ versus } t_D/C_D$$

with each curve defined by a value of

$$\delta[(C_D \exp(2S)_{f+m}/\lambda \exp(-2S)]$$

where

$$\delta = \frac{6}{\gamma^2} = 1{\cdot}89 \qquad \text{for horizontal slab matrix blocks} \qquad (29)$$

$$\delta = \frac{10}{3\gamma^2} = 1{\cdot}05 \qquad \text{for spherical matrix blocks} \qquad (30)$$

($\gamma = 1{\cdot}78$ is the exponential of the Euler constant)

As already mentioned, these families of curves represent two extreme transition cases. Type-curve matching involves three curves, sequentially

—first, one of the $C_D \exp(2S)$ curves of value

$$(C_D \exp(2S))_f = \frac{0{\cdot}8936C \exp(2S)}{(\phi V c_t)_f h r_w^2} \qquad (31)$$

for pressure data corresponding to the response of the most permeable system.

—second, one of the $\lambda \exp(-2S)$ or $\delta[(C_D \exp(2S))_{f+m}/\lambda \exp(-2S)]$ curves for pressure data during transition

—third, another of the $C_D \exp(2S)$ curves with a $C_D \exp(2S)$ value:

$$(C_D \exp(2S))_{f+m} = \frac{0{\cdot}8936C \exp(2S)}{(\phi V c_t)_{f+m} h r_w^2} \qquad (32)$$

smaller than $(C_D \exp(2S))_f$, for pressure data corresponding to the total system.

Because the type-curves in Figs 31 and 32 do not represent exact double-porosity behaviour, but only its various components, the match during transition cannot be perfect, especially when $(C_D \exp(2S))_{f+m}$ is not very small compared to $(C_D \exp(2S))_f$ (high ω values). The transition curve to be used for quantitative analysis is the one passing through the mid-point of the transition zone.

Type-curve analysis of Δp versus Δt yields:

1. The permeability–thickness product kh from the pressure match (eqn (9)); kh represents in fact $k_f h$, the permeability–thickness for the most permeable system (the least permeable system does not produce into the well).
2. The wellbore storage constant C from the time match (eqn (13)).
3. The skin factor S from the curve match with the total system homogeneous curve $(C_D \exp(2S))_{f+m}$ (eqn (32)).
4. ω from the ratio of the $C_D \exp(2S)$ for the total system to that for the most permeable system:

$$\omega = \frac{(C_D \exp(2S))_{f+m}}{(C_D \exp(2S))_f} = \frac{(\phi V c_t)_f}{(\phi V c_t)_{f+m}} \qquad (33)$$

5. λ from the transition curve match ($\lambda \exp(-2S)$) or

$$\delta[(C_D \exp(2S))_{f+m}/\lambda \exp(-2S)]$$

depending upon the nature of interporosity flow.

In the case of unrestricted interporosity flow, the shape of the matrix blocks must be known or must be assumed to obtain λ.

The values $k_f h$, C, S, ω and λ represent the total amount of information that can be extracted from the test in the absence of boundary effects. In addition, it is possible to evaluate the quality of the well from the value of $(C_D \exp(2S))_f$, i.e. from the curve match for the most permeable system. In general, damaged wells correspond to $(C_D \exp(2S))_f$ values greater than 1000; non-damaged wells, to $(C_D \exp(2S))_f$ values between 5 and 1000; acidised wells to $(C_D \exp(2S))_f$ values between 0·5 and 5; and fractured wells to $(C_D \exp(2S))_f$ values less than 0·5.

If the well is non-damaged, it is also possible to distinguish between fissured and multilayered reservoirs. Non-damaged or stimulated wells in fissured reservoirs exhibit wellbore storage values from log–log or cartesian analysis that are 10 to 100 times higher than those computed from completion data, because most of the storage effect is due to the fissure volume. On the other hand, the same in multilayered reservoirs exhibit C values that are in line with those computed from completion.[2]

Conversely, a high wellbore storage constant associated with a very negative skin value is indicative of a fissured reservoir even when the test only shows a homogeneous behaviour. This would occur when the test is too short, so that only the most permeable system can be seen.

ASSOCIATED SPECIALISED PLOTS: Limits of the various flow regimes for the homogeneous components of the double porosity behaviour are the same as in the type-curve for homogeneous reservoir behaviour (Fig. 24). Consequently, storage flow exists only for damaged wells, i.e. for $(C_D \exp(2S))_f$ greater than 1000. The start of infinite acting radial flow on each $C_D \exp(2S)$ homogeneous curve is indicated by a continuous curve in Figs 31 and 32. Actual pressure points matching a $C_D \exp(2S)$ homogeneous curve beyond this line are located on a semi-log straight line on the semi-log specialised plot. If this occurs for both the $(C_D \exp(2S))_f$ and the $(C_D \exp(2S))_{f+m}$ curves, two semi-log straight lines are present on the specialised plot. These two straight lines are parallel because they both relate to the permeability of the most permeable system. In the case of unrestricted interporosity flow (Fig. 32), a third semi-log straight line may exist if actual pressure points during transition match a transition curve outside the region limited by two dotted curves in Fig. 32. This transition semi-log straight line has a slope equal to half that of the semi-log straight lines corresponding to infinite acting radial flow.

4.2.3. Superposition Type-Curves[1]

The drawdown type-curves presented in the previous sections are not strictly valid for analysing data of flow periods (whether drawdowns or build-ups), after the first drawdown. For such flow periods, the change in pressure for log–log analysis is defined as (Fig. 33):

$$\Delta p = p(\Delta t) - p(\Delta t = 0) \tag{34}$$

The corresponding theoretical type-curve is obtained from the appropriate drawdown type-curve as

$$(p_D)_{Bu} = p_D[(\Delta t)_D] + p_D[(t_p)_D] - p_D[(t_p + \Delta t)_D] \tag{35}$$

in the case of a build-up flow period, and as

$$(p_D)_{MR} = \sum_{i=1}^{n-1} \frac{q_i - q_{i-1}}{q_{n-1} - q_n} \left\{ p_D\left[\sum_{j=1}^{n-1} (\Delta t_j)_D\right] - p_D\left[\left(\sum_{j=1}^{n-1} \Delta t_j + \Delta t\right)_D\right]\right\} + p_D(\Delta t)_D \tag{36}$$

in the case of a multiple rate test. In eqn (35) t_p represents an equivalent production time prior to shut-in. In eqn (36), the Δt_j values and the q values represent the duration of the various flow periods and the average rates during these flow periods, respectively; n is the number of the flow periods being analysed. The build-up type-curve from eqn (35) and the multirate type-curve from eqn (36) must be evaluated for each test, because they depend upon the test conditions.

In practice, drawdown type-curves can still be used for type-curve analysis of build-up or multirate data, provided that:[1]

$$p_D[(t_p)_D] - p_D[(t_p + \Delta t)_D] \quad \text{in eqn (35)}$$

or each

$$p_D\left[\sum_{j=1}^{n-1} (\Delta t_j)_D\right] - p_D\left[\left(\sum_{j=1}^{n-1} \Delta t_j + \Delta t\right)_D\right] \quad \text{in eqn (36)}$$

is negligible. This occurs when Δt is small compared to t_p or to $\sum \Delta t_j$ (the duration of the flow period under consideration is small compared to the cumulative duration of all previous flow periods in the test), or when the above differences are small compared to $p_D[(\Delta t)_D]$ (the drawdown type-

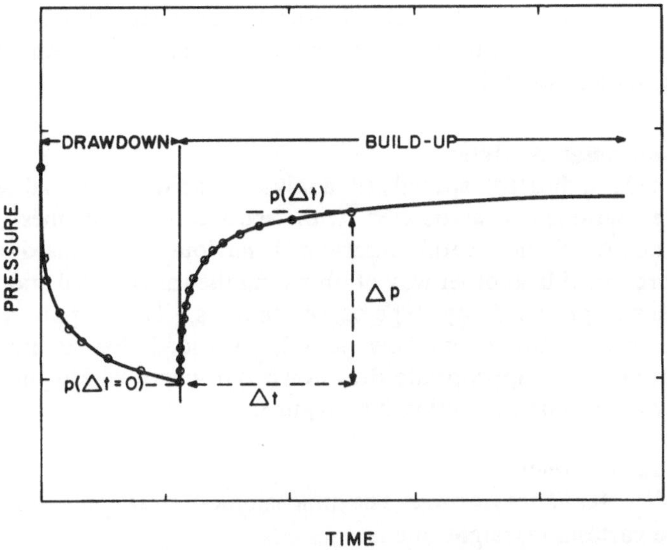

FIG. 33. Build-up pressure test.

curve exhibits large values, usually in the presence of high skin). Under such conditions, actual pressure data plotted on a log–log graph as Δp (from eqn (34)) versus Δt do match a drawdown type-curve. However, they start deviating from it as soon as the above conditions no longer hold true, which occurs at higher values of Δt.

Type-curve analysis can still be performed in the usual way (except that q must be replaced by $\Delta q = q_{n-1} - q_n$ in eqn (9)), provided that it is possible to identify the correct drawdown type-curve without ambiguity. This is sometimes difficult, especially when the durations of the previous flow periods are small compared to the duration of the flow period under study. In such a case, it is usually necessary to generate build-up or multirate type-curves, as appropriate.

Specialised plots, on the other hand, may or may not be applicable and reveal straight lines where they would be expected. Correct straight lines for the various flow regimes exist only through the actual pressure points that match a drawdown type-curve on the log–log plot. When the actual pressure points deviate from a drawdown type-curve and follow a build-up or multirate type-curve, specialised analyses become invalid, even though in some cases these points may appear to be located on a straight line on the appropriate specialised plot: such a straight line is not the correct one and would yield inconsistent results if used for analysis. This has been the subject of much confusion in the literature, especially with respect to infinite acting radial flow.

4.3. Specialised Analyses

As already indicated, specialised analyses, based on specialised plots, provide parameters associated with specific flow regimes and are independent of the overall reservoir behaviour. Specialised analyses, therefore, provide another way of obtaining the various well and reservoir parameters provided by type-curve analysis. They apply to initial drawdowns or subsequent flow periods, provided that actual pressure points match the appropriate drawdown type-curve within the range of validity of the corresponding flow regime.

4.3.1. Early Times

4.3.1.1. Wellbore Storage. The dimensionless equation of the wellbore storage cartesian straight line is:[3]

$$p_D = t_D/C_D \tag{37}$$

Specialised analysis for wellbore storage flow is based on the specialised plot Δp versus Δt shown in Fig. 7, and yields the wellbore storage constant C:

$$C \text{ (in bbl/psi)} = \frac{\Delta q B}{24 m_{ws}} \tag{38}$$

where m_{ws} is the slope of the wellbore storage cartesian straight line in Fig. 7.

4.3.1.2. High conductivity vertical fracture. The dimensionless equation of the infinite conductivity fracture cartesian straight line is[4]

$$p_D = \sqrt{(\pi t_{Df})} \tag{39}$$

Specialised analysis for high conductivity vertical fracture flow is based on the specialised plot Δp versus $\sqrt{\Delta t}$ shown in Fig. 9, and yields the product kx_f^2 of the reservoir permeability by the square of the fracture half-length:

$$kx_f^2 \text{ (in mD ft}^2\text{)} = \frac{16{\cdot}52\mu}{\phi c_t}\left(\frac{\Delta q B}{h m_{IKVF}}\right)^2 \tag{40}$$

where m_{IKVF} is the slope of the infinite conductivity vertical fracture straight line in Fig. 7.

4.3.1.3. Low conductivity vertical fracture. The dimensionless equation of the finite conductivity fracture cartesian straight line is[5]

$$p_D = \frac{2 \cdot 45}{\sqrt{(k_{fD} w_{fD})}} \sqrt[4]{t_{Df}} \quad (41)$$

Specialised analysis for low conductivity vertical fracture flow is based on the specialised plot Δp versus $\sqrt[4]{\Delta t}$ shown in Fig. 11, and yields the fracture conductivity:

$$k_f w \text{ (in mD ft)} = \frac{1945 \cdot 8 \, (\Delta q B \mu)^2}{(h m_{FKVF})^2 (\phi \mu c_t k)^{1/2}} \quad (42)$$

where m_{FKVF} is the slope of the finite conductivity vertical fracture straight line in Fig. 11.

4.3.2. Infinite Acting

The dimensionless equation of the infinite acting semi-log straight line is

$$p_D = \tfrac{1}{2}(\ln t_{De} + 0 \cdot 809\,07) \quad (43)$$

Specialised analysis for infinite acting radial flow is known in the literature as Miller Dyes and Hutchinson (MDH)[16] analysis and is based on the specialised plot Δp versus $\log \Delta t$ shown in Fig. 13. It yields:

1. The permeability–thickness product:

$$kh \text{ (in mD ft)} = 162 \cdot 6 \frac{\Delta q B \mu}{m} \quad (44)$$

2. The skin factor:

$$S = 1 \cdot 151 \left[\frac{\Delta p(\Delta t = 1\,\text{h})}{m} - \log \frac{k}{\phi \mu c_t r_w^2} + 3 \cdot 23 \right] \quad (45)$$

where m is the slope of the infinite acting radial flow straight line in Fig. 13, and $\Delta p(\Delta t = 1\,\text{h})$, the Δp value of the point on this straight line corresponding to $\Delta t = 1$ h.

Specialised analysis for infinite acting radial flow applies to the pressure points matching an homogeneous type-curve within the range of validity of infinite acting radial flow. In the case of heterogeneous behaviour, there may be two infinite acting radial flow straight lines on the semi-log specialised plot: specialised analysis must be applied to the semi-log straight line representing the total system, with c_t in eqn (45) replaced by $(\phi V c_t)_{f+m}$. If the characteristic two parallel semi-log straight lines exist in the case of double-porosity behaviour (a rare occurrence, as already mentioned), the distance δp between the straight lines can be used to calculate (Fig. 34):

$$\omega = 10^{-\delta p/m} \tag{46}$$

4.3.3. Late Times

The dimensionless equation of the pseudo-steady-state cartesian straight line is[9]

$$p_D = 2\pi t_{DA} + (p_{int})_D \tag{47}$$

where t_{DA} is the dimensionless time based on the drainage area A:

$$t_{DA} = t_{De} \times \frac{r_{we}^2}{A} \tag{48}$$

and $(p_{int})_D$ is the dimensionless intercept, a function of the drainage area shape factor C_A:

$$(p_{int})_D = \frac{1}{2}\left[-\ln \frac{r_{we}^2}{A} - \ln C_A + 0{\cdot}809\,07\right] \tag{49}$$

Specialised analysis for closed boundary effects (also called pseudo-steady-state analysis, or reservoir limit test analysis) is based on the specialised cartesian plot of Δp versus Δt shown in Fig. 17. It yields:

1. The area of the closed reservoir:

$$A \text{ (in ft}^2) = \frac{0{\cdot}234 \Delta q B}{\phi c_t h m_{PSS}} \tag{50}$$

2. The reservoir shape factor:[17]

$$C_A = 5{\cdot}456 \frac{m}{m_{PSS}} 10^{[\Delta p(\Delta t = 1\text{ h}) - \Delta p_{int}/m]} \tag{51}$$

where m_{PSS} is the slope of the pseudo-steady-state cartesian straight line in Fig. 17; Δp_{int} is the intercept of this straight line with the Δp-axis; and m and $\Delta p(\Delta t = 1\text{ h})$ are the slope and intercept with $\Delta t = 1$ h of the infinite acting semi-log straight line, respectively. The value of the shape factor C_A is related to the shape of the drainage area.

4.4. Horner Type-Curves and Horner Analysis

Infinite acting data from flow periods after the first drawdown can be also analysed in terms of

$$\Delta p = p_i - p(\Delta t) \tag{52}$$

versus

$$\frac{t_p + \Delta t}{\Delta t} \tag{53}$$

in the case of a build-up, or versus

$$\sum_{i=1}^{n-1} (q_i - q_{i-1}) \log\left(\sum_{j=1}^{n-1} \Delta t_j + \Delta t\right) + (q_n - q_{n-1}) \log \Delta t \tag{54}$$

in the case of a multirate test. t_p, q_i, Δt_j and n in eqns (53) and (54) have the same meaning as in eqns (35) and (36).

In practice, $p(\Delta t)$ is used instead of eqn (52) because p_i is usually not available. In fact, p_i or the average reservoir pressure is one of the parameters obtained from the analysis, in addition to the other parameters available from infinite acting data, i.e. kh, S and λ. Analysis involves a semi-log plot of

$$p(\Delta t) \text{ versus } \frac{t_p + \Delta t}{\Delta t}$$

in the case of a multirate test . t_p, q_i, Δt_j and n in eqns (53) and (54) have the mathematical expression (eqn (54)) in the case of a multirate test.[18] The former is called a Horner plot, and the corresponding analysis method is known as Horner analysis. The latter is called a superposition plot (or generalised Horner plot).

4.4.1. Build-Up Test

Figure 34 represents a Horner plot. A characteristic of a Horner plot is that pressure points during infinite acting radial flow are located on a semi-log straight line. The straight line extrapolates to p^* (called the extrapolated pressure, or the 'false' pressure) at infinite shut-in times:

$$\frac{t_p + \Delta t}{\Delta t} = 1$$

p^* is equal to the initial pressure p_i in the case of an undepleted reservoir, and can be used to calculate the average reservoir pressure in the case of a depleted reservoir.

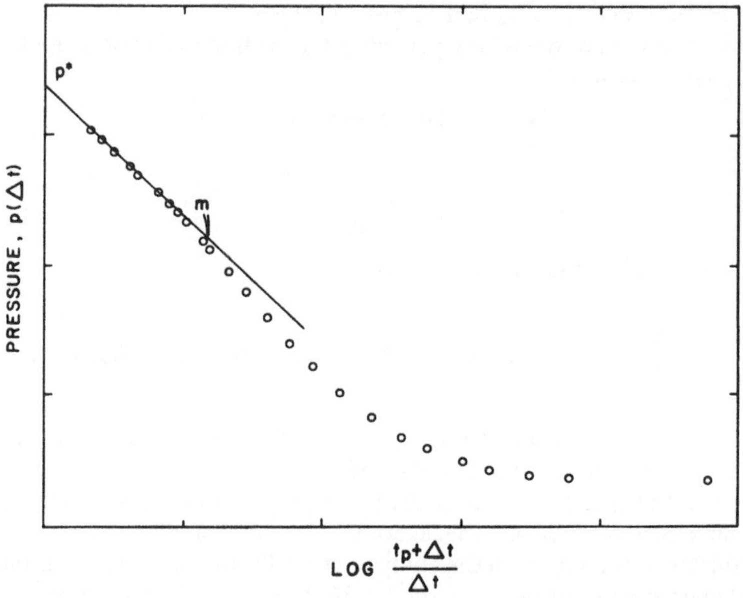

FIG. 34. Horner plot.

Analysis yields:

1. The permeability–thickness product:

$$kh \text{ (in mD ft)} = 162{\cdot}6 \frac{\Delta q B \mu}{m} \tag{55}$$

2. The skin factor:

$$S = 1{\cdot}151 \left[\frac{p(\Delta t = 1\,\text{h}) - p(\Delta t = 0)}{m} - \log \frac{k}{\phi \mu c_t r_w^2} + \log \frac{t_p + 1}{t_p} + 3{\cdot}23 \right] \tag{56}$$

3. The extrapolated pressure, p^*

where m is the slope of the Horner semi-log straight line in Fig. 34 and $p(\Delta t = 1\text{ h})$, the pressure value of the point on this straight line corresponding to $\Delta t = 1$ h $[(t_p + \Delta t)/\Delta t = t_p + 1]$.

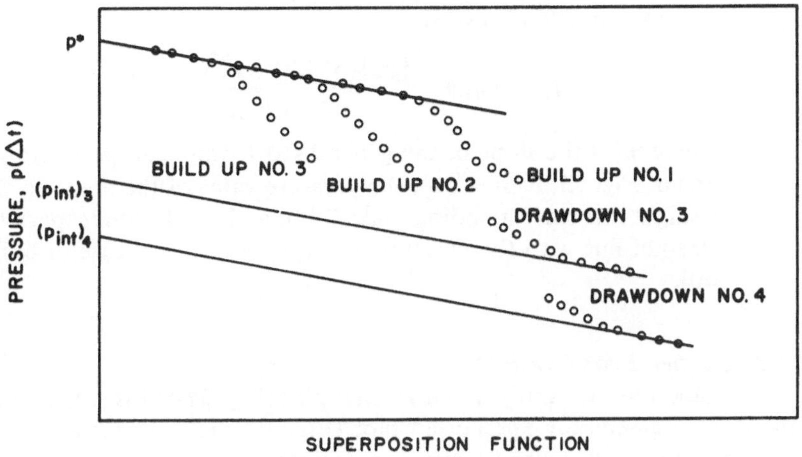

FIG. 35. Generalised Horner plot.

4.4.2. Multirate Test

Figure 35 presents a generalised Horner plot for various flow periods of the same test. A characteristic of this plot is that pressure points during infinite acting radial flow are located on a cartesian straight line. This straight line is the same for all individual build-up periods in the test, and extrapolates to p^* when expression (54) is equal to zero. Straight lines for individual drawdowns, on the other hand, are all different, but are parallel to the build-up straight line. Analysis yields:

1. The permeability thickness product:

$$kh \text{ (in mD ft)} = 162 \cdot 6 \frac{B\mu}{m'} \tag{57}$$

2. The skin factor:

$$S = 1 \cdot 151 \left[\frac{p(\Delta t = 1\text{ h}) - p(\Delta t = 0)}{m'(q_n - q_{n-1})} - \log \frac{k}{\phi \mu c_t r_w^2} + 3 \cdot 23 + \sum_{i=1}^{n-1} \frac{q_i - q_{i-1}}{q_{n-1} - q_n} \log \frac{\sum_{j=i}^{n-1} \Delta t_j + \Delta t}{\sum_{j=i}^{n-1} \Delta t_j} \right] \tag{58}$$

3. The extrapolated pressure:

$$p^* = p_{\text{int}} + \frac{q_n[p(\Delta t = 1\text{ h}) - p(\Delta t = 0)]}{q_{n-1} - q_n} \qquad (59)$$

where m' is the slope of the generalised Horner cartesian straight line in Fig. 35; $p(\Delta t = 1\text{ h})$, the pressure value of the point on this straight line corresponding to $\Delta t = 1$ h; and p_{int} the intercept of the straight line with the pressure axis ($p_{\text{int}} = p^*$ in the case of build-up).

4.4.3. Horner Type Curves[19]

Two considerable advantages of a Horner plot (Fig. 34 or Fig. 35) over an infinite acting semi-log specialised plot (Fig. 13) are that: (1) a Horner straight line may exist when a semi-log straight line on the specialised plot does not; and (2) when both straight lines exist, the Horner straight line is likely to be of longer duration. The reason is as follows:

The dimensionless type-curve corresponding to Horner analysis is given as

$$p_{D_H} = \frac{kh}{141{\cdot}2\Delta q B\mu}[p_i - p(\Delta t)] = p_D(t_p + \Delta t_D) - p_D(\Delta t_D) \qquad (60)$$

versus

$$\frac{t_p + \Delta t}{\Delta t}$$

For p_{D_H} to be a semi-log function of $(t_p + \Delta t)/\Delta t$, it is only necessary that $p_D(\Delta t)_D$ be a semi-log function of Δt, because $p_D(t_p + \Delta t)_D$ will automatically be one, unless outer boundary effects are overwhelming. In other words, it is only necessary that Δt be large enough so that infinite acting homogeneous radial flow occurs on the drawdown type-curve, but actual pressure points on the log–log plot do not need to match the drawdown type-curve.

On the other hand, an infinite acting semi-log straight line can only exist on the infinite acting semi-log specialised plot if the build-up type-curve

$$p_{DBU} = \frac{kh}{141{\cdot}2\,\Delta q B\mu}[p(\Delta t) - p(\Delta t = 0)]$$
$$= p_D(\Delta t)_D + p_D(t_p)_D - p_D(t_p + \Delta t_D) \qquad (36)$$

versus $(\Delta t)_D$:

(1) can be approximated by the drawdown type-curve $p_D(\Delta t)_D$ *and*
(2) infinite acting homogeneous radial flow occurs on the drawdown type-curve. In other words, it is necessary that actual pressure points on the log–log plot match the drawdown type-curve beyond the start of infinite acting homogeneous behaviour.

The same comments apply in the case of a multirate test, for which the dimensionless type-curve corresponding to generalised Horner analysis is given as

$$p_{D_{GH}} = \frac{kh}{141{\cdot}2\,\Delta q B\mu}[p_i - p(\Delta t = 0)]$$

$$= \sum_{i=1}^{n-1} \frac{q_i - q_{i-1}}{q_{n-1} - q_n} p_D\left[\left(\sum_{j=i}^{n-1} \Delta t_j + \Delta t\right)_D\right] - p_D(\Delta t)_D \qquad (61)$$

versus

$$\sum_{i=1}^{n-1} \frac{q_i - q_{i-1}}{q_{n-1} - q_n} \log\left[\sum_{j=i}^{n=1} \Delta t_j + \Delta t\right] - \log \Delta t \qquad (62)$$

For the case of reservoirs with heterogeneous behaviour, there may be several appropriate straight lines on a Horner, or generalised Horner plot, corresponding to infinite acting homogeneous behaviour of each component of the system. Analysis must be performed with the straight line corresponding to the infinite acting homogeneous behaviour of the total system, with ϕc_t being replaced by $(\phi V c_t)_{f+m}$ in eqns (56) and (58). This straight line, however, only exists if the duration of the preceding flow period is such that total system behaviour has been reached. Figure 36 illustrates several possibilities that can be found in practice, in the case of a double-porosity reservoir with restricted interporosity flow. Two sets of build-up data are shown on a Horner plot, each corresponding to a different drawdown duration.[2]

Curve E corresponds to a drawdown ending after the total system homogeneous curve has been reached. It exhibits a complete double-porosity behaviour, with, in this case, two Horner straight lines. The second straight line, corresponding to the total system, is the one to be used for analysis. The distance between the two straight lines can be used to calculate ω from eqn (46). Curve A corresponds to a drawdown ending on the most permeable system homogeneous curve, before the start of transition. It only exhibits homogeneous behaviour. The Horner straight line corresponds to the most permeable system. Curves B, C and D correspond to drawdowns ended during the transition period. As a result, none of the features expected from double-porosity behaviour develop, and Horner analysis cannot be performed. Worse, if this fact is not recognised, a straight line could be found on the plot and analysed as a Horner straight line, thus leading to erroneous results, and possibly to decisions that are

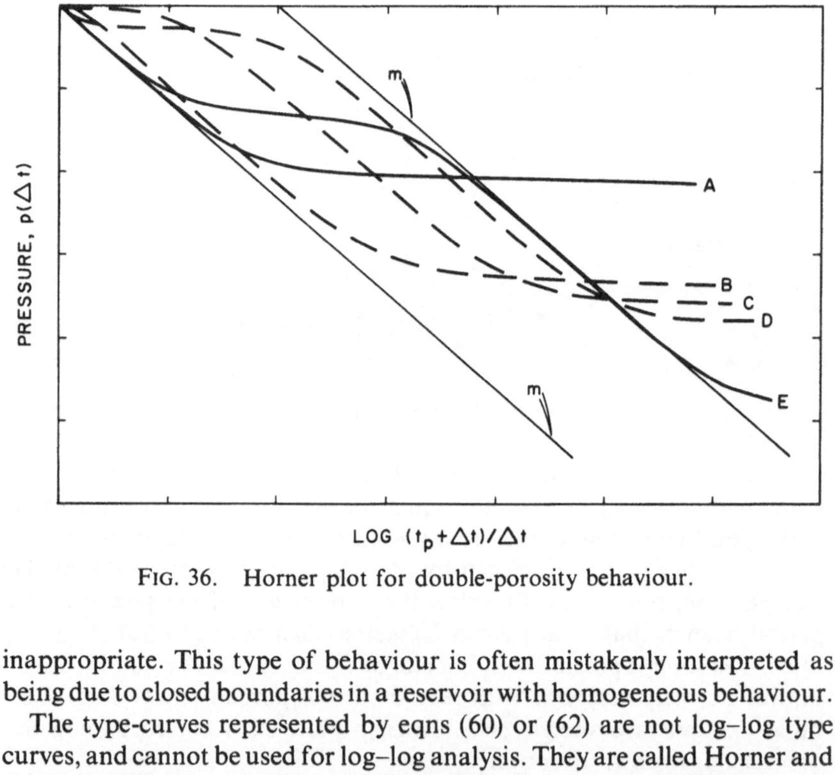

FIG. 36. Horner plot for double-porosity behaviour.

inappropriate. This type of behaviour is often mistakenly interpreted as being due to closed boundaries in a reservoir with homogeneous behaviour.

The type-curves represented by eqns (60) or (62) are not log–log type curves, and cannot be used for log–log analysis. They are called Horner and generalised Horner type-curves, respectively. As in the case of log–log type curves, they exhibit shapes that are characteristic of flow regimes and reservoir behaviour, but emphasise these characteristic shapes for time ranges different from those on log–log plots. They are mainly used for checking interpretation results. Horner type-curves exhibit a straight line of slope 1·151 during infinite acting homogeneous behaviour.

5. MODEL VERIFICATION

The final stage of well test interpretation is a check of its validity. Three techniques can be used.

5.1. Consistency Check

As discussed throughout this chapter, several methods are used to obtain the model parameters: (1) type-curve analysis provides all the parameters pertinent to the model selected for analysis; (2) specialised analysis provides parameters specific to individual flow regimes; and (3) Horner analysis provides parameters specific to infinite acting behaviour.

When the interpretation is correct, these different analysis techniques must yield the same numerical value for a given parameter. In practice, some difference (e.g. 10% for kh) is acceptable. Unrealistic differences indicate interpretation inconsistency, which may be due to an erroneous interpretation model, or inconsistent or missing data.

5.2. Dimensionless Horner Plot[19]

A dimensionless Horner plot can be used to check the validity of the analysis. This is achieved by plotting independently on the same graph:

(1) the Horner, or generalised Horner, type-curve corresponding to the interpretation model selected for analysis, and

(2) the actual pressure data for the flow period being analysed, plotted in dimensionless Horner, or generalised Horner form, as

$$\frac{kh}{141{\cdot}2\,\Delta qB\mu}[p_\mathrm{i}-p(\Delta t=0)]$$

versus

$$\frac{t_\mathrm{p}+\Delta t}{\Delta t}$$

for a build-up or

$$\sum_{i=1}^{n-1}\frac{q_i-q_{i-1}}{q_{n-1}-q_n}\log\left(\sum_{j=i}^{n=1}\Delta t_j+\Delta t\right)-\log\Delta t \qquad (62)$$

for a multirate test, respectively. The various parameters needed to generate these two curves (i.e. kh, p_i, time match and curve match) are the ones obtained from the analysis. For the analysis to be consistent, the Horner type-curve and the dimensionless Horner actual data should match. This type of plot is more sensitive to errors than a log–log graph, and it is possible that a reasonable log–log match yields an inconsistent dimensionless Horner match. An unsatisfactory match, however, can be used to adjust interpretation results.

Figure 37 illustrates an error in the evaluation of the initial pressure p_i. This error results in actual data being parallel to the Horner type-curve,

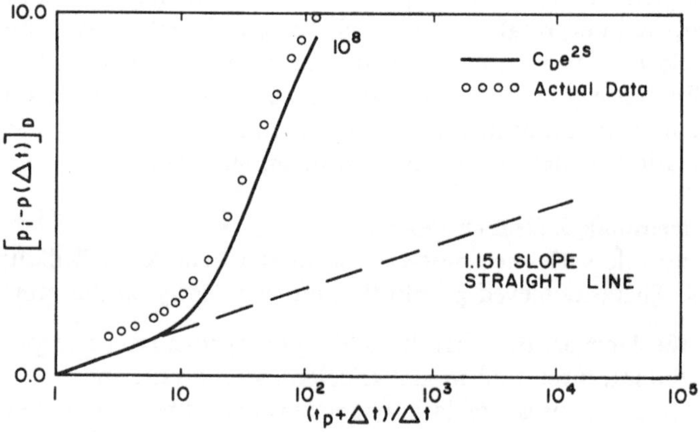

FIG. 37. Dimensionless Horner plot = erroneous p_i.

and shifted by an amount proportional to the error in p_i, which can then be evaluated.

Figure 38 illustrates an incorrect selection of the Horner straight line. Dimensionless actual data do exhibit a 1·151 slope Horner straight line, but they do not match the type-curve.

Alternatively, the dimensionless Horner plot can be helpful to interpret a test when data are insufficient to apply without ambiguity the various techniques described before. Figure 39 shows various log–log matches with the same set of build-up data, that are equally acceptable. Results from log–log analysis, however, are widely different in terms of kh and skin.

If a value for p_i is available from other sources, a dimensionless Horner plot can be prepared for each match shown in Fig. 39. Erroneous log–log matches become obvious on such a plot. For instance, Fig. 40 indicates that

$C_D \exp(2S) = 10^4$ is too high a curve match value, whereas Fig. 41 indicates that 10^2 is too low. The most consistent curve match can then be obtained by trial and error.

Another interesting feature of dimensionless Horner plotting is illustrated in Fig. 42: shapes for homogeneous and heterogeneous behaviour are very different, whereas it could be difficult to distinguish them on a log–log plot.

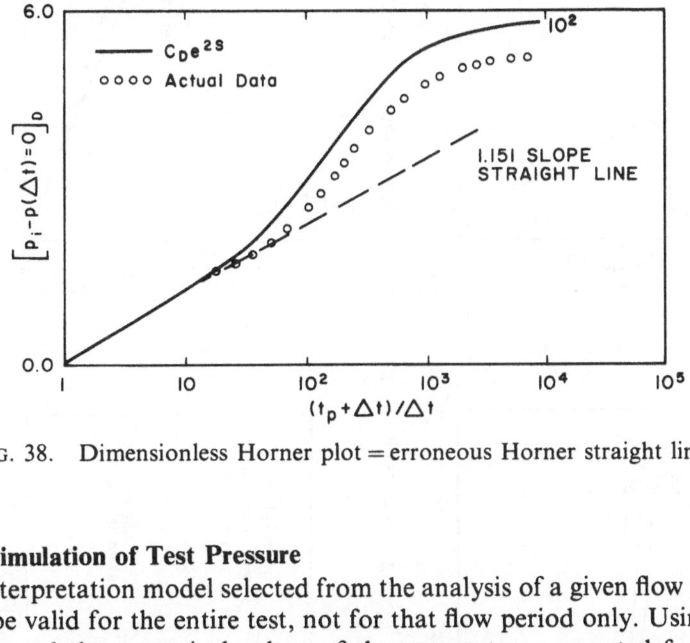

FIG. 38. Dimensionless Horner plot = erroneous Horner straight line.

5.3. Simulation of Test Pressure

The interpretation model selected from the analysis of a given flow period must be valid for the entire test, not for that flow period only. Using this model, and the numerical values of the parameters computed from the analysis, it is possible to simulate the test pressure history, taking into account the actual rate history, and to compare it with the actual pressure history. Again, any inconsistency would indicate an erroneous interpretation.

This procedure is particularly useful to reveal a diagnostic error when an infinite reservoir with double-porosity behaviour has been mistakenly analysed as a closed reservoir with homogeneous behaviour. Usually, the size computed for the latter is so small that the simulated depletion during the initial drawdown is much higher than the actual depletion, although the analysis of individual build-ups may appear consistent.

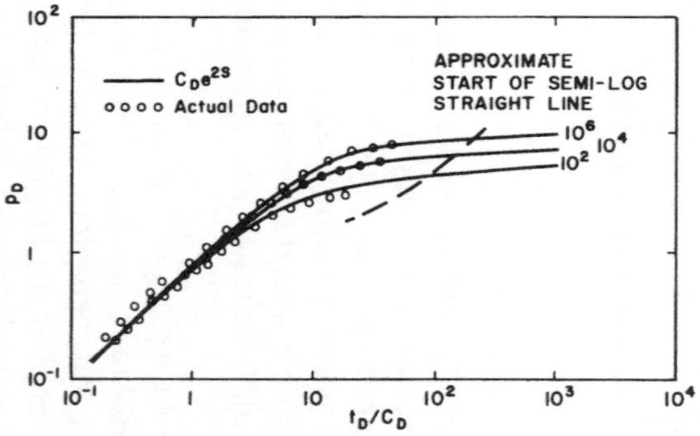

FIG. 39. Example of possible log–log matches with insufficient data.

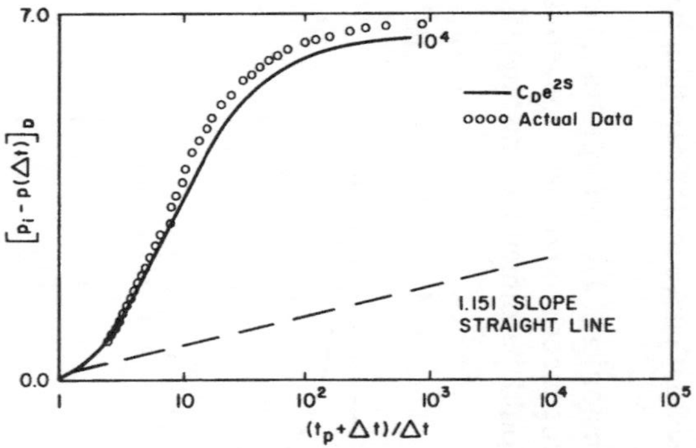

FIG. 40. Dimensionless Horner plot = too high a $C_D \exp(2S)$ value.

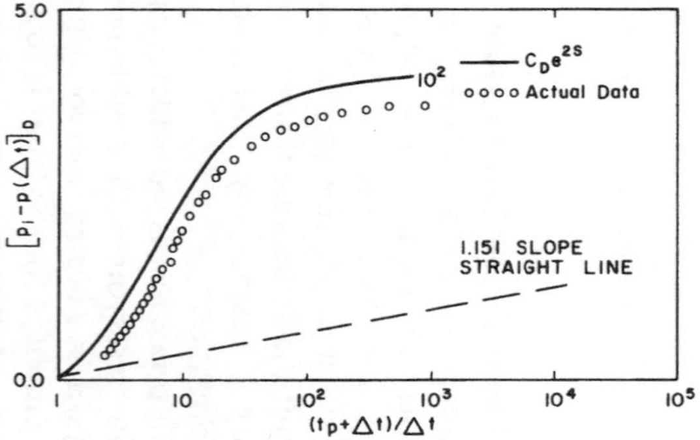

FIG. 41. Dimensionless Horner plot = too low a $C_D \exp(2S)$ value.

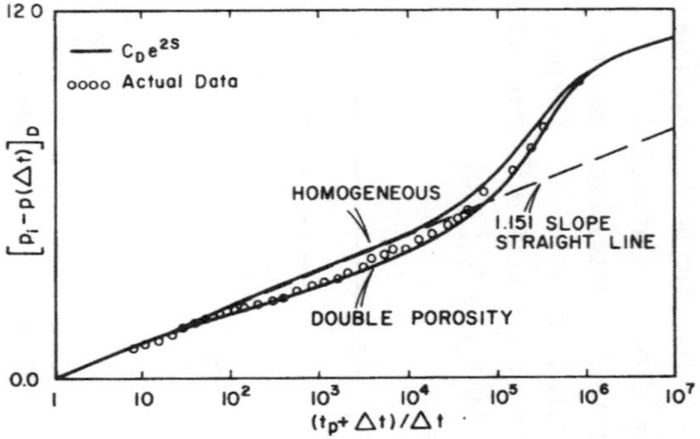

FIG. 42. Dimensionless Horner plot = homogeneous versus double-porosity behaviour.

6. EXAMPLE

The interpretation methodology described in this chapter is illustrated with the test example listed in Table 2 and plotted in Fig. 43. Test data have been computer generated in order to be able to demonstrate as much of the interpretation methodology as possible. Well and reservoir information is

TABLE 2
TEST EXAMPLE

Time (*h*)	*Pressure* (*psia*)	*Time* (*h*)	*Pressure* (*psia*)
40·0	3 146·0	50·0	3 636·9
42·5	3 143·9	50·5	3 645·8
45·0	3 141·8	51·0	3 652·7
47·5	3 139·9	52·0	3 663·2
47·7	3 139·7	54·0	3 677·1
47·9	3 139·6	56·0	3 686·5
End of drawdown		58·0	3 693·3
48·0	3 139·6	60·5	3 699·9
48·003	3 148·1	63·0	3 705·1
48·004	3 150·9	68·0	3 712·7
48·005	3 153·7	70·5	3 715·6
48·006	3 156·4	71·75	3 716·9
48·007	3 159·1	73·0	3 718·1
48·008	3 161·8	74·25	3 719·2
48·009 5	3 165·8	75·5	3 720·3
48·012 5	3 173·7	76·75	3 721·3
48·015	3 180·1	78·0	3 722·3
48·02	3 192·5	80·5	3 724·0
48·025	3 204·4	83·0	3 725·5
48·030	3 215·9	85·5	3 726·9
48·04	3 237·6	88·0	3 728·2
48·06	3 276·7	90·5	3 729·4
48·08	3 310·6	93·0	3 730·4
48·1	3 340·0	95·5	3 731·4
48·15	3 398·5	98·0	3 732·3
48·2	3 441·0	100·5	3 733·1
48·25	3 472·7	103·0	3 733·9
48·3	3 497·1	105·5	3 734·6
48·4	3 531·1	108·0	3 735·3
48·5	3 553·7	113·0	3 736·5
48·6	3 569·6	118·0	3 737·6
48·8	3 590·6	120·0	3 738·0
49·0	3 604·2		

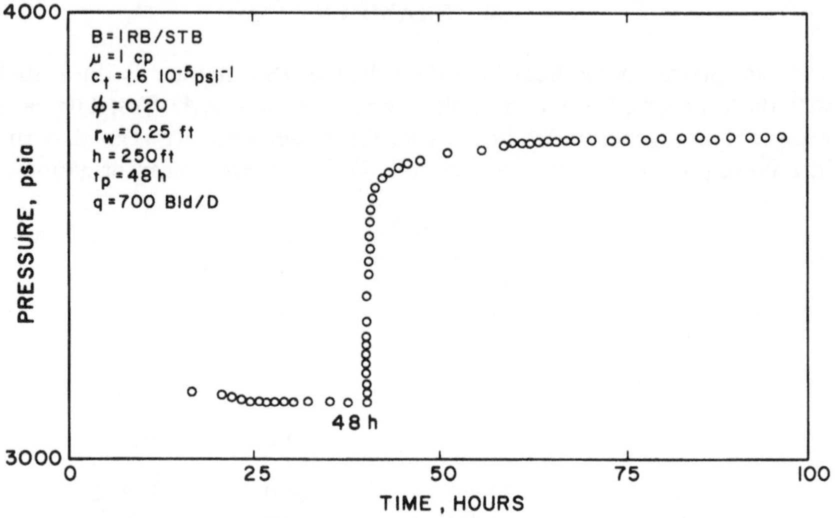

FIG. 43. Well test example: pressure history.

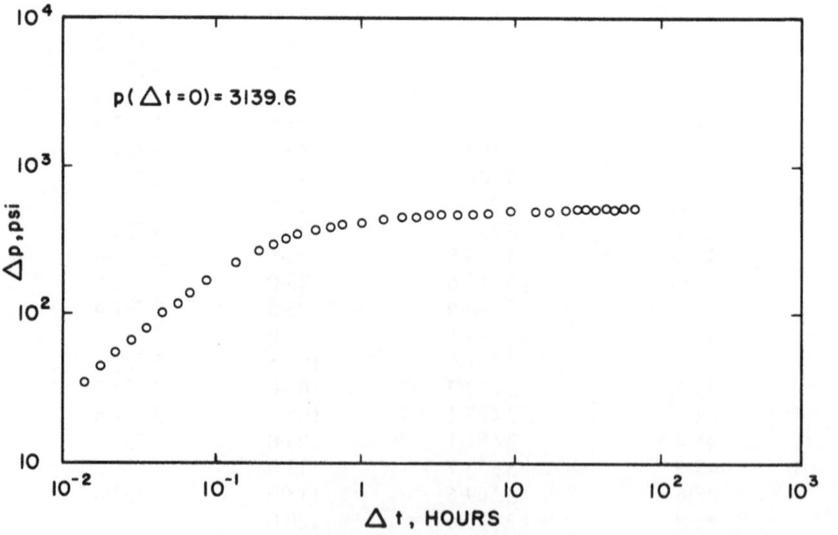

FIG. 44. Well test example: log–log plot.

shown in Fig. 43. The test consists of a 48-h drawdown at a constant rate of 700 bbl/D, followed by a 72h build-up. Analysis is performed on the build-up. As indicated before, the interpretation process starts with a log–log plot of the build-up data $[p(\Delta t) - p(\Delta t = 0)]$ versus Δt, where $p(\Delta t = 0)$ is the pressure at the end of the drawdown, Δt the build-up time, and $p(\Delta t)$ the corresponding pressure reading (Fig. 44).

The shape of the log–log plot in Fig. 44 suggests a well with wellbore storage in a reservoir with homogeneous behaviour. Early time data are located on a unit slope straight line, which indicates a damaged well. Build-up data will thus have to be matched with one of the type-curves of Fig. 24, corresponding to $C_D \exp(2S)$ greater than 1000.

An initial type-curve match is shown in Fig. 45. The purpose of this initial match is to confirm the diagnostic of the model and to check if Horner analysis is applicable. Horner analysis allows evaluation of kh and therefore the determination of the value of the pressure match, thus reducing type-curve matching to one degree of freedom, the time match.

Figure 45 indicates a reasonable match with $C_D \exp(2S) = 10^4$, with Horner analysis valid after 4 h (pointer number 2). It can be seen that the build-up data do not match the drawdown type-curve after approximately 10 h (pointer number 3) which indicates that the semi-log specialised analysis is only valid between 4 and 10 h. The match also indicates that the wellbore storage specialised analysis is only applicable to measurements taken before a Δt of 30 s (pointer number 1).

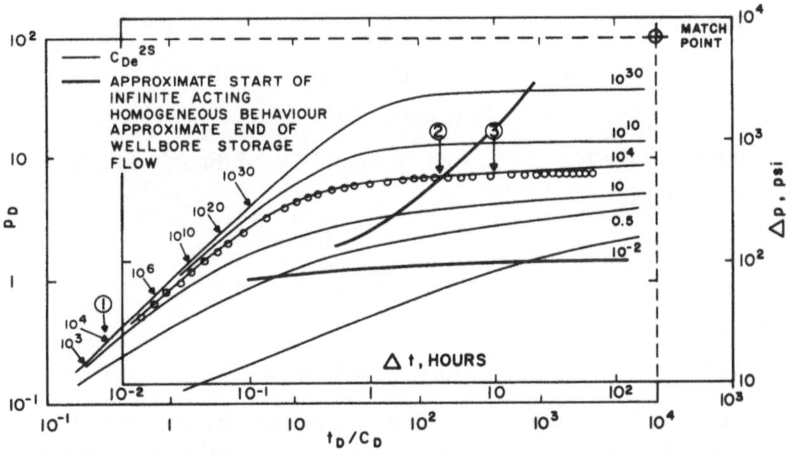

FIG. 45. Well test example: type-curve matching procedure.

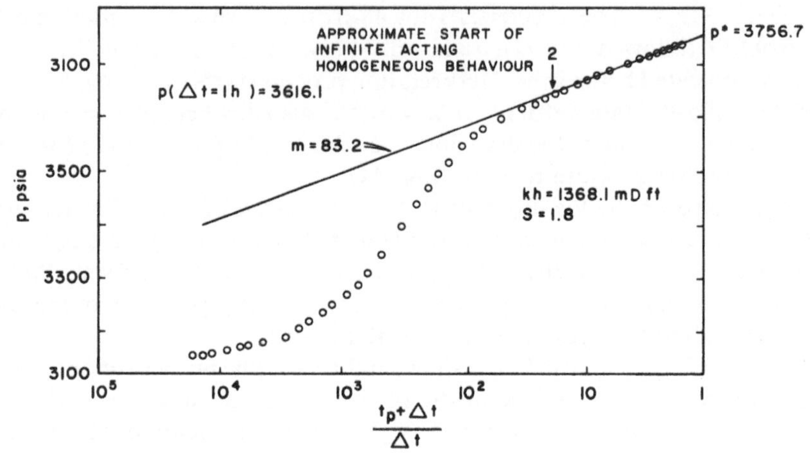

FIG. 46. Well test example: Horner plot.

Figure 46 shows a Horner plot. The Horner straight line starts indeed at 4 h $(t_p + \Delta t)/\Delta t = 13$ as predicted from the type-curve match. The Horner straight line has a slope

$$m = 83{\cdot}2 \text{ psia/log cycle}$$

and an intercept with $(t_p + \Delta t)/\Delta t = 1$ of

$$p^* = 3756{\cdot}7 \text{ psi}$$

This yields:

$$kh = 1368{\cdot}1 \text{ mD ft from eqn (55)}$$

$$S = 1{\cdot}8 \text{ from eqn (56)}$$

The slope m of the Horner straight line can be used directly to calculate the pressure match:

$$(p_D/\Delta p)_{\text{match}} = \frac{1{\cdot}151}{m}$$

which yields

$$(p_D/\Delta p)_{\text{match}} = 0{\cdot}0138$$

A refined type-curve match can then be obtained by matching p_D on the dimensionless type-curve with $\Delta p = 72{\cdot}3$ on the actual pressure graph, and shifting time axes. The pressure match can then be slightly altered to

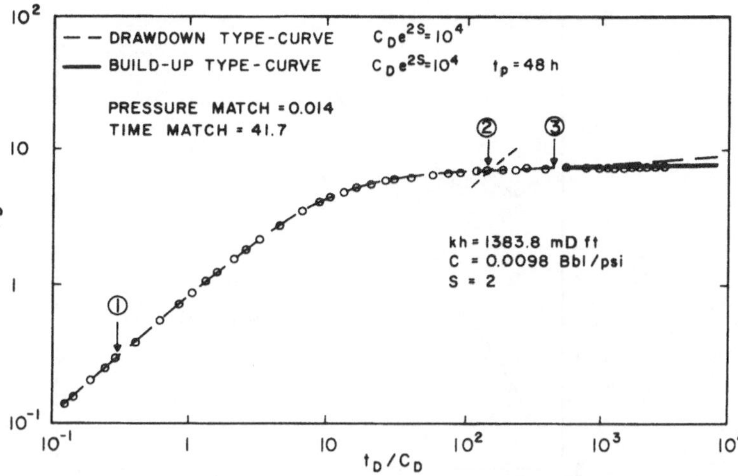

FIG. 47. Well test example: type-curve match.

improve the fit. The result is shown in Fig. 47: a good match is obtained with the build-up type curve corresponding to $C_D \exp(2S) = 10^4$ and $t_p = 48$ h, with a pressure match given as

$$(p_D/\Delta p)_{\text{match}} = 0{\cdot}014 (p_D = 1 \text{ for } \Delta p = 70)$$

and a time match as

$$\left(\frac{t_D/C_D}{\Delta t}\right)_{\text{match}} = 41{\cdot}7 (t_D/C_D = 1 \text{ for } \Delta t = 0{\cdot}024)$$

These yield:

$$kh = 1383{\cdot}8 \text{ mD ft from eqn (9)}$$
$$C = 0{\cdot}009\,8 \text{ bbl/psi from eqn (13)}$$
$$S = 2 \text{ from eqn (14)}$$

The next step in the interpretation process involves specialised analyses, to provide independent estimates of some of the parameters obtained from the type-curve analysis. Two specialised analyses are applicable: namely, that for wellbore storage to data before 30 s, and that for infinite acting homogeneous behaviour (semi-log radial flow) to data between 4 and 10 h.

Figure 48 presents a cartesian specialised plot for wellbore storage

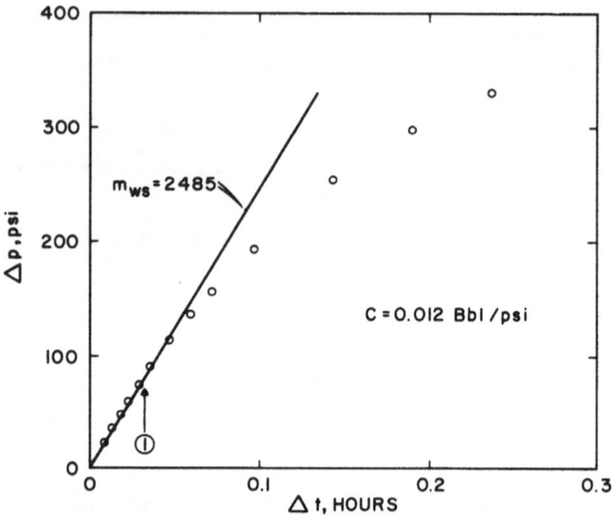

FIG. 48. Well test example: wellbore storage specialised plot.

dominated flow. Early time pressure points are located on a straight line of slope:

$$m_{ws} = 2495 \text{ psi/h}$$

which yields:

$$C = 0{\cdot}012 \text{ bbl/psi from eqn (38)}$$

Figure 49 shows an infinite acting specialised (MDH) plot. As indicated by the log–log match of Fig. 47, the infinite acting semi-log straight line exists through the pressure points between 4 and 10 h, with a slope equal to

$$m = 74{\cdot}74 \text{ psi/log cycle}$$

and an intercept with $\Delta t = 1$ h equal to 479·3. These yield:

$$kh = 1522 \text{ mD ft from eqn (44)}$$
$$S = 2{\cdot}9 \text{ from eqn (45)}$$

To complete the interpretation process, checking procedures are used to check the consistency of the interpretation. Table 3 presents all the results obtained by the various methods and gives the values of the parameters used for computing the test data. All results are consistent, which indicates

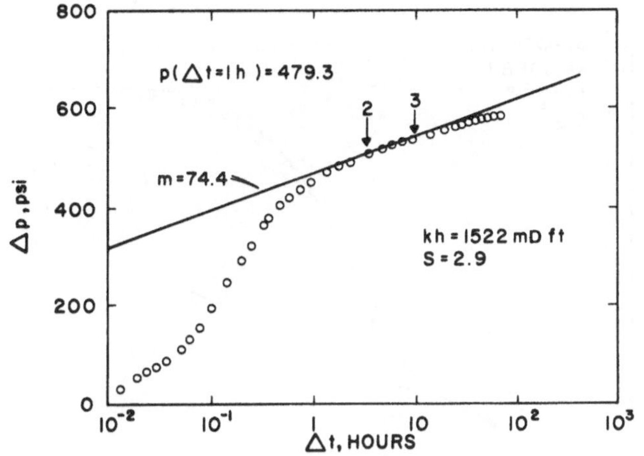

FIG. 49. Well test example: infinite acting specialised plot.

a consistent interpretation; *kh* values are within 10% of one another, and within 10% of the design value.

Figure 50 represents a dimensionless Horner plot, derived with the parameters obtained from Horner analysis. A good match is obtained with the Horner type-curve, indicating again a consistent interpretation.

Finally, Fig. 51 shows a comparison between actual pressure history and that simulated from analysis results. A first simulation using results from Horner analysis gives too high a pressure drop during the drawdown, indicating too low a *kh* value ($kh = 1368$ mD ft), although the match between simulated and actual pressure is very good during the build-up. A simulation with $kh = 1428$ mD ft yields too low a pressure drop during drawdown, suggesting a lower actual *kh* value. A final *kh* value was selected

TABLE 3
CHECK OF INTERPRETATION CONSISTENCY

Analysis method	*kh* (*mD ft*)	*C* (*bbl/psi*)	*S*	*p** (*psia*)
Type-curve analysis	1 383·8	0·009 8	2	
Horner analysis	1 368·1		1·8	3 756·7
Wellbore storage specialised analysis		0·012		
Infinite acting specialised analysis	1 522		2·9	
Design parameters	1 412	0·01	2	3 756

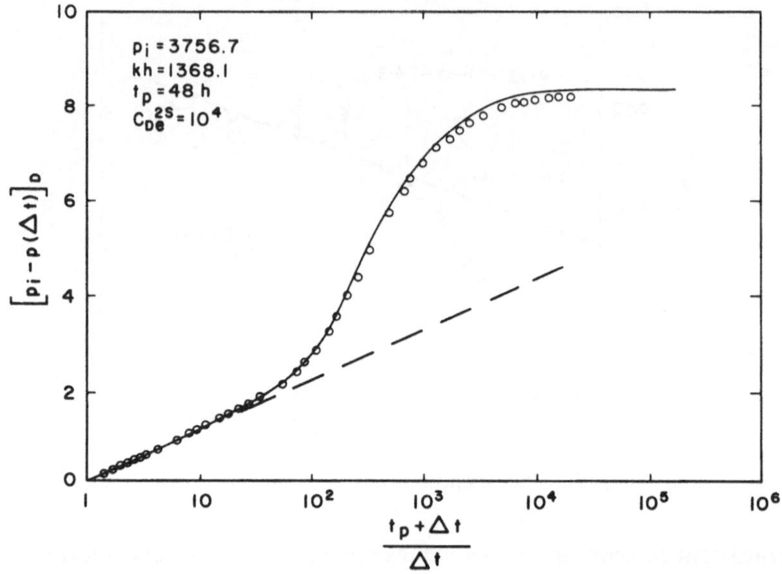

FIG. 50. Well test example: dimensionless Horner plot.

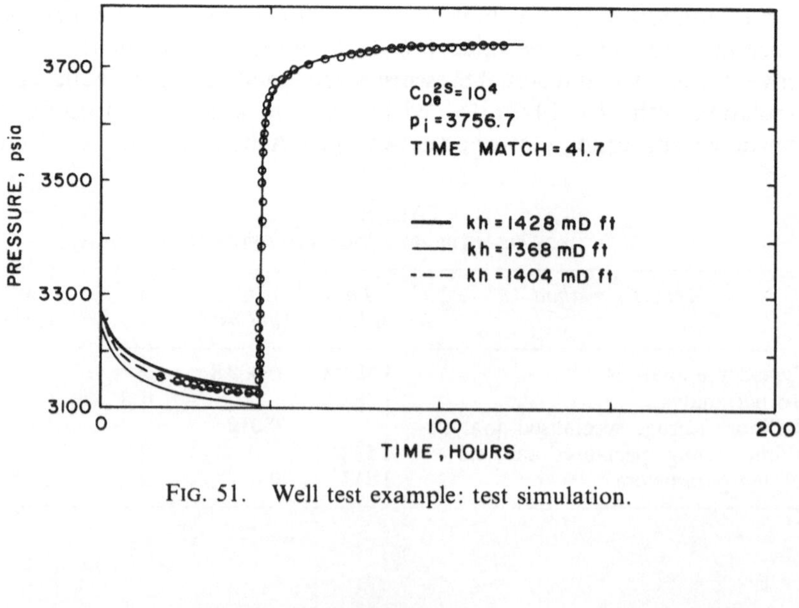

FIG. 51. Well test example: test simulation.

as 1404 mD ft, which yields a good fit with actual drawdown data. This is within 0·5% of the design value of 1412 mD ft.

Based on this kh value of 1404 mD ft the complete interpretation procedure can be repeated to refine the analysis. Final results are:

$$kh = 1404\ \text{mD ft}$$
$$C = 0{\cdot}01\ \text{bbl/ft}$$
$$S = 2$$
$$p^* = 3756{\cdot}7\ \text{psi}$$

7. TEST DESIGN

Well tests are easy to interpret if they are designed properly. Unfortunately, tests are usually not designed with interpretation in mind, and there is a general tendency to mix requirements for production assessment (p versus q) with those for reservoir behaviour evaluation. As a result, choke size is changed repeatedly during the drawdown, which makes interpretation of any flow period very difficult. A good practice is to organise the test so that measurements for reservoir behaviour evaluation are performed first, and those for production evaluation are taken afterwards.

Tests must be designed so that: (1) changes in rate and pressure to be measured correspond to the sensitivity of the gauges to be used; (2) the duration of each flow period during the test is compatible with the reservoir features to be observed; and (3) as many of the interpretation techniques described herein as compatible with the economics of the test can be used to assess the consistency of the interpretation.

In general, a reservoir feature cannot be seen in a build-up if it is not seen while the well is flowing. Therefore, in order to see, during a build-up, boundaries or all the components of a double-porosity behaviour that are required to fully characterise a fissured or multilayered reservoir, it is necessary that the drawdown be long enough for these features to be fully developed before starting the build-up. This is usually the main difficulty of well testing, as the time required for important features to appear is usually not known beforehand. In reservoirs with homogeneous behaviour, it is usually possible to obtain minimum distances to boundaries by default, even if they do not show on test data. Also, a complete build-up analysis is generally possible in infinite reservoirs even with a short preceding drawdown, provided that the drawdown is not entirely dominated by wellbore storage, and that the build-up is long enough for Horner analysis to apply.

In the case of double-porosity behaviour, on the other hand, it is imperative for complete analysis that total system behaviour be reached during *both* drawdown and build-up. If the drawdown is too short, no or only limited double-porosity information can be obtained from the build-up, no matter how long the well is shut-in. Thus, tests in double-porosity reservoirs have to be planned carefully, using information available from previous tests on the same well. However, because fluid and well condition changes may have an important impact on test behaviour in a given well,[20] the most efficient way to ensure adequate test duration is to monitor the bottom hole pressure during the test, using surface read-out with electronic gauges and an electric cable, and to interpret pressure data as described in this chapter while they are being acquired.

8. CONCLUSIONS

The purpose of this chapter was to present a systematic approach to the interpretation of well tests. The methodology described herein is applicable to any kind of test (DST, production or injection), or to any kind of fluid (water, oil or gas). The only difference in the case of gas is that analysis must be performed with the real gas potential function $m(p)$[21] instead of the pressure p:

$$m(p) = \int_{p_0}^{p} \frac{2p\,\mathrm{d}p}{\mu Z}$$

where Z, the real gas derivation factor, and μ, the gas viscosity, are known functions of pressure. Analysis can also be performed in terms of pressure if the pressure during the test remains above 3000 psi, or in terms of pressure squared, if the pressure during the test remains below 1000 psi. The methodology is also applicable to the multiphase flow situation, and yields total mobility instead of kh, when total fluid production is taken into account.

The key characteristic of the interpretation process is that it includes three distinct and successive steps:

identification
calculation
verification

each one using sequentially many of the 'modern' or 'conventional' techniques that have been published in the literature over the years. Identification relies heavily on the log–log analysis method, developed initially in hydrology[14] and introduced later in petroleum engineering.[3]

An improvement, however, over the pioneering work of a decade ago[22,23] is that the type-curves being used are expressed in terms of independent variables, which make them easier to use and more efficient. In particular, it is now possible to differentiate, with little ambiguity, homogeneous from double-porosity behaviour, and to explain phenomena that used to be considered as erratic. More important, because parameters carry different meanings depending upon the reservoir behaviour, better and more profitable decisions can now be made from well test data—the most visible being the ability of selecting the most appropriate completion for the well.

The combination of diagnosis methods and checking procedures, as described herein, make well test analysis a technique that can be learned and applied easily, and usually guarantees the best interpretation possible, given the quality of the data being analysed and the current state of interpretation techniques.

REFERENCES

1. GRINGARTEN, A. C., BOURDET, D., LANDEL, P. A. and KNIAZEFF, V. (1979) A comparison between different skin and wellbore storage type-curves for early-time transient analysis. Paper SPE 8205, *SPE-AIME 54th Annual Fall Technical Conference and Exhibition*, Las Vegas, Nevada.
2. GRINGARTEN, A. C. (1982) Interpretation of tests in fissured reservoirs and multilayered reservoirs with double porosity behavior. Paper SPE 10044, *International Petroleum Exhibition and Technical Symposium of the Society of Petroleum Engineers*, Bejing, China.
3. RAMEY, H. J., JR (1970) Short-time well test data interpretation in the presence of skin effect and wellbore storage. *J. Pet. Tech.* (January), 97.
4. GRINGARTEN, A. C., RAMEY, H. J., JR and RAGHAVAN, R. (1975) Applied pressure analysis for fractured wells. *J. Pet. Tech.* (July), 887–92.
5. CINCO, L. H., SANANIEGO, V. F. and DOMINGUEZ, A. N. (1978) Transient pressure behavior for a well with a finite-conductivity vertical fracture. *Soc. Pet. Eng. J.* (August), 253.
6. VAN EVERDINGEN, A. F. and HURST, W. (1949) The application of the Laplace Transformation to flow problems in reservoirs. *Trans. AIME*, **186**, 305–24.
7. WARREN, J. E. and ROOT, P. J. (1963) Behavior of naturally fractured reservoirs. *Soc. Pet. Eng. J.* (September), 245.

8. Bourdet, D. and Gringarten, A. C. (1980) Determination of fissured volume and block size in fractured reservoirs by type-curve analysis. Paper SPE 9293, *SPE-AIME 55th Annual Fall Technical Conference and Exhibition*, Dallas, Texas.
9. Jones, P. (1956) Reservoir limit tests. *Oil and Gas Journal*, **54**(59), 184.
10. Horner, D. R. (1951) Pressure build-up in wells. *Proc. Third World Pet. Cong.*, E. J. Brill, Leiden, **II**, 503–21.
11. Bourdet, D., Whittle, T. M., Douglas, A. A. and Priard, Y. M. (1983) A new set of type-curves simplifies well test analysis. *World Oil* (May), 95–106.
12. Bourdet, D., Ayoub, J. A., Whittle, T. M., Pirard, Y. M. and Kniazeff, V. (1983) Interpreting well tests in fractured reservoirs. *World Oil* (Oct.), 77–87.
13. Earlougher, R. C., Jr (1977) *Advances in Well Test Analysis*, SPE Monograph Series Number 5, Society of Petroleum Engineers of AIME, Dallas, Texas.
14. Theis, C. F. (1935) The relationship between the lowering of the piezometric surface and the rate and duration of discharge using ground-water storage. *Trans. AGU*, **2**, 519.
15. Gringarten, A. C. (1978) Reservoir limit testing for fractured wells. Paper SPE 7452, *SPE-AIME 53rd Annual Fall Technical Conference and Exhibition*, Houston, Texas.
16. Miller, C. C., Dyes, A. B. and Hutchinson, C. A., Jr (1950) Estimation of permeability and reservoir pressure from bottom-hole pressure build-up characteristics. *Trans. AIME*, **189**, 91–104.
17. Earlaugher, R. C., Jr (1971) Estimating drainage shapes from reservoir limit tests. *J. Pet. Tech.* (October), 1266.
18. Odeh, A. S. and Jones, L. G. (1965) Pressure drawdown analysis, variable rate case. *J. Pet. Tech.* (August), 960.
19. Gringarten, A. C., Bourdet, D., Fjaere, D. S. and Viturat, D. (1980) Horner type-curve analysis. Paper SPE 8291, *SPE-AIME 55th Annual Fall Technical Conference and Exhibition*, Dallas, Texas.
20. Gringarten, A. C., Burgess, T. M., Viturat, D., Pelissier, J. and Aubry, M. (1981) Evaluating fissured formation geometry from well test data: a field example. Paper SPE 10182, *SPE-AIME 56th Annual Fall Technical Conference and Exhibition*, San Antonio, Texas.
21. Al-Hussainy, R. and Ramey, H. J., Jr (1966) Application of real gas flow theory to well testing and deliverability forecasting. *J. Pet. Tech.* (May), 637–42.
22. Agarwal, R. G., Al-Hussainy, R. and Ramey, H. J., Jr (1970) An investigation of wellbore storage and skin effect in unsteady liquid flow. I: Analytical treatment. *Soc. Pet. Eng. J.* (September), 279.
23. McKinley, R. M. (1971) Wellbore transmissibility from afterflow-dominated pressure buildup data. *J. Pet. Tech.* (July).

Chapter 6

RESERVOIR MODELLING: FINITE ELEMENTS

JOHN RAE
AERE Harwell, Oxon, UK

1. DIFFICULTIES IN MODELLING AND HOW FINITE ELEMENTS CAN HELP

In reservoir simulation we mimic the behaviour of fluids in the porous rock of reservoirs by mathematical equations based on three types of physical laws: the principle of conservation of mass, or material balance; equations of fluid motion, usually the Darcy flow equations; and thermodynamic or PVT properties of the fluids concerned. In all but the simplest problems the equations take the form of a number of coupled non-linear partial differential equations which are much too difficult to solve other than by numerical methods. The traditional procedure is to divide up the reservoir into a large number of cells which have a mathematical meaning but only a loose connection with the physical or geological description of the reservoir. The rock and fluid properties of these cells are presumed known and the job of the numerical part of the modelling is to calculate the flow from each cell to the others and to and from the wells which operate in the reservoir. Mathematically this is equivalent to making a finite difference representation of the basic modelling equations and is described in many texts, for example Refs 1–3, and in other chapters of this book.

A number of difficulties may arise both from geometrical and numerical aspects of this approach:

(1) The reservoir may have curved boundaries. If the simulator program has a facility for flagging out inactive cells this may not matter, as a rough representation of the boundary is usually adequate. If a better representation is needed one may have to resort to complicated mapping techniques.[4]

(2) The boundary conditions may be complicated. Fixed pressure values or fixed flow values generally cause no problem but a boundary through which a solute flows may have quite complicated boundary conditions.[5] It may be difficult to treat these with enough accuracy.

(3) There may be special parts of the reservoir where the flow is extra complicated, for example near wells or faults. Well models go some way towards handling this problem,[6] but the adequacy of their representation is often questionable especially when many wells cluster together or are completed away from cell centres.

(4) It is difficult to use different cell meshes in different parts of the reservoir. This may apply to fine meshes in the regions of interest, such as near wells, or to coarse meshes in regions of little interest such as aquifers. The theory and practice of finite difference methods on very irregular meshes are poor and normally a fine mesh placed on one part has to be propagated across much of the reservoir where it is not needed with consequent increases in computer costs.

(5) The mathematical model equations take the form of advection–diffusion equations; material is moved around as a Darcy flow and spreads by capillary action, hydrodynamic dispersion and molecular diffusion. It is well known that these equations have stability problems when the advection dominates.[3,7] The only ways round these problems in conventional simulators are to use higher-order methods or to refine the mesh or to use some form of upstreaming technique. Higher-order methods are difficult to program and have never been popular although attempts have been made.[8] Mesh refinement of the whole reservoir to an adequate degree is extremely costly even for the modern highly effective linear equation solvers. Selective mesh refinement, particularly with moving meshes, has been successful in one-dimensional simulations[9] but its cost–benefit ratio rises sharply in two or three dimensions. Upstreaming methods although popular and very common, lead to false diffusion effects which may destroy the structure of the flow being studied. A mesh refinement to control this leads back to cost problems.

The problems listed have been around since the beginning of computational fluid mechanics and much work has gone, and is going, into their solution. No panacea has been found or is likely to be. Finite element methods, applied in fluid mechanics since the early 1970s,[10–12] can help in certain respects.

(1) Curved boundaries are easily handled.
(2) Complex boundary conditions can usually be handled easily.
(3) Special elements can be placed at chosen positions in the mesh to model effects of wells or faults or other special features. One can even mix regions with completely different physical behaviours, such as a channel with Navier–Stokes flow in a permeable rock with Darcy flow.
(4) A mesh of finite elements can have elements of almost any size, almost any shape and use many different shape functions (this is the polynomial interpolation used on elements, and it is described later). Mesh refinements can be done in two ways, either by introducing more elements locally or by using higher-order shape functions on those elements.[13]
(5) The finite element method has a natural matrix equation solver, the frontal solution method,[14,15] which takes special account of the structure of finite elements and, at least in two-dimensional problems, is superbly efficient.[16] It is a direct solver whose reliability appeals to engineers but is nonetheless as fast as anything generally available.

None of the difficulties mentioned above are impossible to overcome in the finite difference method but sometimes the finite element method handles them much more naturally. The most important advantages are that the mesh can be refined and moulded to the problem in essentially arbitrary ways and that high-order shape functions, which are roughly equivalent to higher-order finite difference methods, can be introduced extremely easily and naturally. Something of a false conflict has grown up between the two methods. In fact there should be no difficulty in reconciling them because, in a sense explained later, the finite element method includes most finite difference schemes as special cases.

2. WHAT IS THE FINITE ELEMENT METHOD?

The first modern applications of the finite element method were to stress analysis of large elastic structures. The engineers who pioneered the method thought of the elements as the actual physical objects, girders, plates, and so on, from which the structures were built. The equations which they were solving in static problems were in fact to minimise the elastic potential energy of the structure. By the mid 1960s mathematicians had realised that this very effective method used in this way was equivalent to the well-known Rayleigh–Ritz variational method.[17]

This realisation led to finite elements being applied to many more problems such as heat flow, potential flow of inviscid fluids, groundwater flow in porous media and creeping viscous flow;[18] all of these examples are cases where a natural variational principle is possible for the problem. In some other cases where there is no natural variational principle, such as flow governed by the Navier–Stokes equations, applications of the finite element method were held up for some years. It was eventually realised that other approaches were possible which were entirely equivalent to a variational principle when there was one, but in fact were more general.[19] Two methods of this type are:

(i) The least-squares method, which despite some drawbacks, is still occasionally used.
(ii) The Galerkin method, which uses the so-called weak form of the equation and is now the most commonly used approach.

These two methods are best illustrated by example and for this we solve the very simple second-order differential equation

$$\frac{d^2u}{dx^2} - 2 = 0 \tag{1}$$

with x in the range $[0, 1]$ and boundary conditions

$$u(0) = 0 \tag{2}$$

$$\frac{du}{dx}(1) = 0 \tag{3}$$

If we look for solutions $u(x)$ which are functions with two derivatives then the usual methods of ordinary differential equations show that there is just the one solution

$$u(x) = x^2 - 2x \tag{4}$$

In the least-squares method the idea is to square eqn (1), integrate it over the range of the variable, and minimise the resulting functional. Thus, we try to find functions $u(x)$ which minimise J given by

$$J = \int_0^1 \left(\frac{d^2u}{dx^2} - 2\right)^2 dx \tag{5}$$

If we succeed in getting J to zero we have solved eqn (1) and the smaller J is, the better our approximate $u(x)$. Note, however, that we have not yet satisfied the boundary conditions. While it is common in all weighted residual methods to choose trial functions which satisfy prescribed value boundary conditions, such as eqn (2), it would be very cumbersome in general and sometimes impossible to arrange that they satisfy all boundary conditions. Thus to satisfy boundary condition (3) we would need to modify the functional J by adding suitable extra terms. This kind of complication with derivative boundary conditions is frequently found with the least-squares approach. Note, also, that the trial functions $u(x)$ we put into J need to be rather smooth. They need two derivatives; so, for example, piecewise linear approximations would not suffice.

For the Galerkin method we proceed differently. Any function $f(x)$ which has at least one derivative and satisfies $f(0) = 0$ we call a test function. If we multiply the left-hand side of eqn (1) by such an f, integrate, and use boundary condition (3), we obtain

$$\int_0^1 \left(\frac{\mathrm{d}^2u(x)}{\mathrm{d}x^2} - 2\right) f(x)\,\mathrm{d}x = \frac{\mathrm{d}u}{\mathrm{d}x} f\Big|_0^1 - 2xf\Big|_0^1 - \int_0^1 \left(\frac{\mathrm{d}u}{\mathrm{d}x} - 2x\right)\frac{\mathrm{d}f}{\mathrm{d}x}\,\mathrm{d}x$$

$$= -\int_0^1 \left(\frac{\mathrm{d}u}{\mathrm{d}x} - 2x + 2\right)\frac{\mathrm{d}f}{\mathrm{d}x}\,\mathrm{d}x \qquad (6)$$

It follows that any solution of our eqns (1)–(3) is also a solution of the set of equations

$$\int_0^1 \left(\frac{\mathrm{d}u}{\mathrm{d}x} - 2x + 2\right)\frac{\mathrm{d}f}{\mathrm{d}x}\,\mathrm{d}x = 0 \qquad (7)$$

for *any* choice of test function f. We may therefore try to solve our original problem by looking for solutions of eqn (7) for enough choices of f. It is not obvious that $u(x) = x^2 - 2x$ is the only solution of eqn (7) satisfying $u(0) = 0$. There could be rather discontinuous solutions of eqn (7) which make no sense for the original eqn (1), and, in fact, this situation does turn up in practice for hyperbolic equations which have solutions with shocks. There is also the question of whether solutions of eqn (7) satisfy the boundary conditions. As before, for our trial functions for u we will use only functions which satisfy $u(0) = 0$, but by choosing test functions f localised at $x = 1$ it turns out that the other boundary condition (3) is automatically satisfied by solutions of eqn (7). Equation (7) is called a *weak* formulation of the original problem; when the same kind of functions are used for the trial functions $u(x)$ and the test functions $f(x)$ we have the

Galerkin method. Although at this stage we may appear to have complicated our problem it should be noted that the Galerkin method has two advantages over least-squares method, namely:

(i) It involves lower-order derivatives than the original equation.
(ii) It incorporates some boundary conditions (*natural* boundary conditions) automatically.

From this point on we will only pursue the Galerkin method.

In practice we proceed as follows. The functions u and f appearing in eqn (7) are now approximated by some chosen set of functions which we will call basis functions. This idea is not new; in fact the Galerkin approach is just what is used in solving linear problems by Fourier analysis or in terms of standard orthogonal polynomials. For example, one might try to solve eqn (1) in terms of the eigenfunctions of the problem

$$\frac{d^2u}{dx^2} = 0 \tag{8}$$

$$u(0) = 0 \tag{9}$$

$$\frac{du}{dx}(1) = 0 \tag{10}$$

which are easily seen to be

$$\sin (2n+1)\frac{\pi x}{2} \qquad n = 0, 1, \ldots \tag{11}$$

These are our basis functions. We expand $u(x)$ in terms of them:

$$u(x) = \sum_{n=0}^{\infty} u_n \sin (2n+1)\frac{\pi x}{2} \tag{12}$$

and the constant 2 in terms of them:[20]

$$2 = \sum_{n=0}^{\infty} \frac{8}{(2n+1)\pi} \sin (2n+1)\frac{\pi x}{2} \tag{13}$$

and put these into eqn (1). Next, we multiply by each basis function in turn (the analogue of the test functions) and integrate. Because the set (11) is orthogonal, this gives us, as usual, equations identifying coefficients of corresponding terms

$$u_n = \frac{32}{(2n+1)^3\pi^3} \tag{14}$$

In terms of the basis functions (11) the solution is then eqn (12) with coefficients given by eqn (14). By chopping off the infinite series at some appropriate stage we could turn this into a numerical scheme for solving eqn (1). In usual applications of the Galerkin finite element method we do not use orthogonal functions, but instead choose basis functions which are piecewise low-order polynomials. Again these basis functions are not new—they were used by Courant in 1943.[21] The simplest form of basis function in one dimension is the piecewise linear or 'hat' function shown in Fig. 1. Its use in the current problem will now be considered.

Continuing with eqn (7), we will now solve it in a finite element approximation. The interval (0, 1) is first broken into sub-intervals, the elements, which we take to be

$$I_1 = (0, \tfrac{1}{4}) \qquad I_2 = (\tfrac{1}{4}, \tfrac{1}{2}) \qquad I_3 = (\tfrac{1}{2}, \tfrac{3}{4}) \qquad I_4 = (\tfrac{3}{4}, 1)$$

The end points of the elements we will call nodes so that we have five nodes and four elements. With each node, except $x = 0$, where we want $u(0) = 0$, we associate a basis function $\phi_i(x)$. The way in which this is done is described in the next section: here we simply state that the first basis function is

$$\phi_1(x) = \begin{cases} 4x & \text{on } I_1 \\ 2 - 4x & \text{on } I_2 \\ 0 & \text{elsewhere} \end{cases} \tag{15}$$

ϕ_2, ϕ_3, ϕ_4 are defined similarly but are shifted by distances of $\frac{1}{4}$, $\frac{1}{2}$ and $\frac{3}{4}$ on the interval (0, 1).

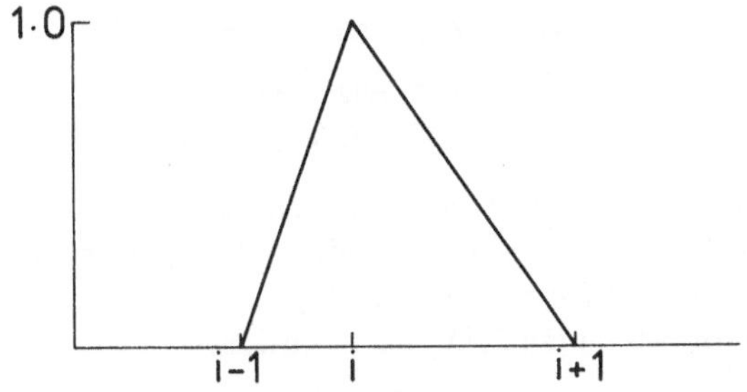

FIG. 1. The one-dimensional linear shape functions.

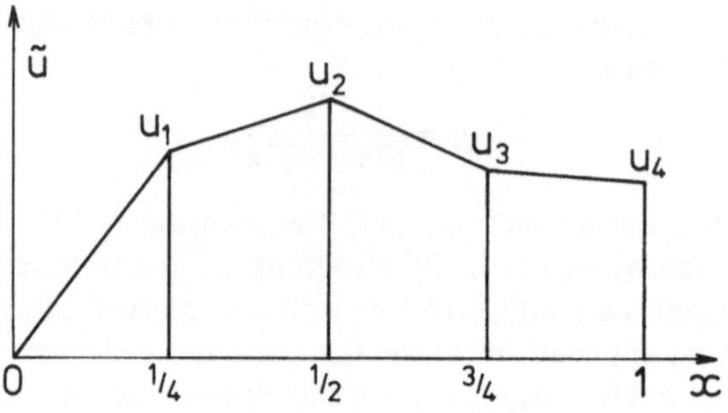

FIG. 2. The form of the approximation to the solution of eqn (7), with u_1–u_4 to be determined.

Our approximation to $u(x)$ and $f(x)$ is then

$$\tilde{u}(x) = \sum_{i=1}^{4} u_i \phi_i(x) \tag{16}$$

$$\tilde{f}(x) = \sum_{i=1}^{4} f_i \phi_i(x) \tag{17}$$

These forms for $\tilde{u}$ and $\tilde{f}$ are put into eqn (7). In asking that the equation hold for all test functions $\tilde{f}$ we are, in this approximation, asking that it hold for all choices of f_i where $i = 1$–4. This is equivalent to asking that it hold when $\tilde{f}$ is replaced by each ϕ_1 in turn. Since the only x dependence in eqn (7) is now in the known functions $\phi_i(x)$ the integrations can be done to give the set of four equations

$$\begin{aligned} 2u_1 - u_2 \qquad\qquad + \tfrac{1}{8} &= 0 \\ -u_1 + 2u_2 - u_3 \qquad + \tfrac{1}{8} &= 0 \\ - u_2 + 2u_3 - u_4 + \tfrac{1}{8} &= 0 \\ - u_3 + u_4 + \tfrac{1}{16} &= 0 \end{aligned} \tag{18}$$

The basis function ϕ_i have been chosen such that the coefficient u_i is actually the value $\tilde{u}(x)$ takes at the node i. Thus our approximation to $u(x)$ at this stage is of the form shown in Fig. 2, where the u_i values are to be determined from the eqns (18). The solution is

$$u_1 = -\tfrac{7}{16} \qquad u_2 = -\tfrac{3}{4} \qquad u_3 = -\tfrac{15}{16} \qquad u_4 = -1$$

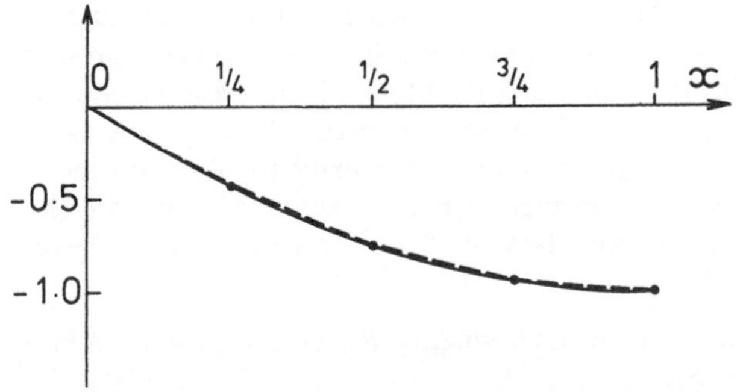

FIG. 3. The finite element approximation to the solution of eqn (7) (dashed line) compared to the exact solution (continuous line).

and our finite element approximation to $u(x)$ is shown in Fig. 3. Because of special features of eqn (1) the nodal values of $\tilde{u}(x)$ are exactly right: in general, this is not true of finite element approximations.

In this very simple example we obtain a fairly good approximation using only four elements. We have not as yet discussed systematically the relationships between elements, nodes and basis functions. The next section will go through the method step by step as it is used in more realistic cases. Also, we have not discussed at all the sense in which our solution is approximate and whether it will improve if we have more elements. These questions lead into difficult technical mathematics and the reader is referred to textbooks such as Ref. 22, or specialist articles such as Ref. 23.

3. THE STEPS OF THE FINITE ELEMENT METHOD IN PRACTICE

The way in which the finite element method is used in real problems is best introduced by an example. The illustration used in the last section was rather too simple for this purpose and a full set of reservoir equations, say for three-phase three-dimensional flow, would be rather too complicated. We choose the advection–diffusion equation for a saturation S

$$\frac{\partial S}{\partial t} + \mathbf{u} \,.\, \nabla f(S) - \nabla \,.\, D\, \nabla S = 0 \tag{19}$$

where t is time, the vector velocity $\mathbf{u} = (u, v)$ is a known function of two-dimensional position x, y, ∇ is the two-dimensional vector gradient operator and the 'dispersion coefficient' D is a known function of position and of the saturation S, as it will be when capillary pressure effects are present. The function f of S might, in oil applications, be a fractional flow. Equation (19), which arises in many flow problems including oil simulations, has just enough complexity to illustrate the steps used in applying the finite element method. Again, to illustrate the method we suppose that the boundary conditions to be applied on the reservoir R are of the following restricted kind:

(a) On a part of the boundary R_1 the saturation is a known function $S = S_1$.
(b) On the rest of the boundary R_2 there is a 'no flux' condition in which the normal derivative of S is zero.

More general boundary conditions can easily be handled[11,12] but would unnecessarily complicate the argument here.

We now go through the steps of the argument in order. This involves some repetition of points of the previous section but is done more fully and systematically here.

Step 1

Step 1 of the development is based on the idea that if eqn (19) is satisfied by some S then so is the equation obtained by multiplying eqn (19) by any test function $h(x, y)$ and integrating over the reservoir region R. Conversely, if this held for enough choices of h one might expect the original eqn (19) to be satisfied. In fact, for equations with dispersion- or diffusion-like terms this is generally true. For equations without such terms, that is hyperbolic equations, shock solutions may develop. The pointwise equation analogous to eqn (19) then needs care in interpretation and the integrated form is the more natural to use. We will touch on this difficulty again later. Now multiply eqn (19) by $h(x, y)$ and integrate over R. We introduce the notation

$$\langle f, g\rangle = \iint_R f(x,y)g(x,y)\,\mathrm{d}x\,\mathrm{d}y \tag{20}$$

and integrate the dispersion term by parts to obtain a boundary integral and shift the gradient operator to the h. Thus

$$\langle \nabla . D\nabla S, h\rangle = \int_{R_1+R_2} D.\frac{\partial S}{\partial n}h\,\mathrm{d}S - \iint_R D\nabla S . \nabla h\,\mathrm{d}x\,\mathrm{d}y \tag{21}$$

where we have used Green's theorem to obtain the normal derivative of S in the boundary integral. The boundary integral is zero on part R_2 of the boundary because $\partial S/\partial n$ is zero there. As in the previous section we restrict our choice of test functions h to those which vanish on R_1, the part of the boundary where S is specified. Boundary terms now disappear and the integrated form, also called the weak or variational form, of eqn (19) is

$$\left\langle \frac{\partial S}{\partial t}, h \right\rangle + \left\langle u\frac{\partial f}{\partial x}, h \right\rangle + \left\langle v\frac{\partial f}{\partial y}, h \right\rangle + \left\langle D\frac{\partial S}{\partial x}, \frac{\partial h}{\partial x} \right\rangle + \left\langle D\frac{\partial S}{\partial y}, \frac{\partial h}{\partial y} \right\rangle = 0 \tag{22}$$

To summarise, we are saying that if eqn (19) and its boundary conditions hold, then so does eqn (22); if eqn (22) holds for 'enough' choices of functions h which vanish on R_1, then so does eqn (19) together with its boundary conditions.

Step 2

At step 2 the approximations begin. The functions appearing in the equations are going to be approximated by linear combinations of a finite number of 'basis' functions. As indicated previously these building blocks can be chosen in many ways. In some simple problems it is obvious that Fourier series or some set of orthogonal polynomials are the best candidates, but we have in mind cases when the geometry and/or the equations are too complex for this. There is no reason in principle why the same basis functions should be used to approximate the unknowns of the equations, such as saturations, and the test functions and in many variants of the method of weighted residuals this is not done.[19] In applying the finite element method, however, the most common approach is that named after Galerkin. This uses the *same* basis functions for both and is the line adopted here. In this context the finite element method can be regarded as a way of choosing convenient sets of basis functions. The main ideas are as follows:

The two-dimensional region R on which eqn (22) applies is divided into pieces, usually triangles or quadrilaterals, which are the finite elements. It is best for now to think of these as having straight sides, although this can be relaxed, but otherwise their number and shape can be chosen freely. A number of special points called nodes are associated with each element. They are usually on the boundaries of elements and are common nodes for all the elements sharing that boundary. For example, vertices of triangles can be taken as nodes and then in Fig. 5 (p. 217) there would be 200 triangular elements and 121 nodes of which only 81 are internal to the region. Quadrilateral elements do not cope well with a curved boundary of this kind but we could, if we had wished, have mixed in some of them in the interior part of the mesh. With each node i we associate a basis function ϕ_i which is a low-order polynomial in each of the elements which share the node and identically zero in all other elements. In the simplest methods of forming the polynomial the number of its coefficients is taken as equal to the number of nodes in the appropriate element and for a polynomial of degree k in any variable we arrange to have $k + 1$ nodes on each edge of the element. Thus, $a + bx + cy$ is a suitable polynomial for the triangles considered above and $a + bx + cy + dxy$ for quadrilaterals. The final step is to choose the coefficients such that, within an element, the polynomial for ϕ_i is unity at the defining node i and zero at all other nodes. This fixes the polynomial uniquely. Functions such as a saturation $S(x, y)$ are now represented in the form

$$S(x, y) = \sum_{\text{nodes}} S_i \phi_i(x, y) \tag{23}$$

From the construction of ϕ_i it is clear that S_i is (the approximation to) the nodal value of S. When the S_i are known, or solved for in the equations, the value of S at any point x, y is given in terms of the ϕ_i. The description given above applies only to the simplest finite elements. Many families of finite elements have been invented for several geometrical shapes (in one-, two- or three-dimensions), different numbers of nodes and different orders of interpolation or shape functions. The reader can find them catalogued in textbooks.[11,24]

At last we can return to the effect of step 2 on eqn (22). There are two further points to get clear. First, eqn (22) is to hold for any choice of test function h, which is to be approximated in the manner of eqn (23):

$$h(x, y) = \sum_{\text{nodes}} h_i \phi_i(x, y) \tag{24}$$

This means eqn (22) is to hold for all choices of the coefficients h_i. But h enters into eqn (22) linearly so it is enough if eqn (22) holds for h replaced by each ϕ_i in turn. If it holds for each separately it holds for any linear combination. Secondly, we should remember that test functions were supposed to take the value zero on the part R_1 of the boundary where S is specified. This is achieved automatically by eqn (24) if we exclude from the choices of i those nodes which lie on R_1. Basis functions for other nodes all go to zero there. In this way eqn (22) is approximated by

$$\sum_i \frac{\partial S_i}{\partial t} \langle \phi_i, \phi_j \rangle + \left\langle u \frac{\partial f}{\partial x} \left(\sum_i S_i \phi_i \right), \phi_j \right\rangle + \left\langle v \frac{\partial f}{\partial y} \left(\sum_i S_i \phi_i \right), \phi_j \right\rangle$$

$$+ \sum_i S_i \left\langle D \frac{\partial \phi_i}{\partial x}, \frac{\partial \phi_j}{\partial x} \right\rangle + \sum_i S_i \left\langle D \frac{\partial \phi_i}{\partial y}, \frac{\partial \phi_j}{\partial y} \right\rangle = 0 \qquad (25)$$

for each node j not on R_1, say for N of these altogether. Each of the sums $\sum_i$ is over all nodes: the known values of S_i for i on R_1 are used when they appear, so that we have the same number N of unknown S_i as there are equations.

This is an important stage of the development, so let us review what we have achieved. The partial differential eqn (19), in its equivalent form (eqn (22)), has been *discretised*—that is represented by the finite set of equations (eqn (25))—a necessary step for solution by computer. The unknowns in eqn (25) are the nodal values S_i which have not been specified in boundary conditions. Note that time-dependence remains in the S_i and that through f and possibly D the equations are nonlinear in the S_i. Apart from this nonlinearity the terms inside the angled brackets are all known functions of x, y and so give numbers when integrated. Thus, eqn (25) is a set of coupled nonlinear first-order differential equations for the S_i as functions of time.

The integrals over x, y are generally done by numerical integration; so, strictly, the choice of integration is a part of the approximation process. In addition, it is important in dealing with elements with curved sides. The trick here is to map on to the curved element in usual x, y coordinates a standard straight-sided element in some suitable local coordinates. The mapping functions are usually approximations, for example in terms of the element basis functions again, which deform the standard element to get its nodes moved to the right places and get reasonable approximations to the curves joining these. This change of coordinates gives rise to extra Jacobian factors in the integrals and numerical integration can handle these without difficulty. Detailed discussion of methods for curved elements, though of great importance in practice, is ingenious and rather lengthy and best left for the textbooks.[11,24]

Step 3

Step 3 deals with the time dependence of eqn (25). We can be brief here for, although it is possible to discretise the time dependence by the finite element method, this is almost never done and conventional time differencing is used. The usual questions arise: whether to use an implicit or explicit method; what order of time difference to use; how to choose time step sizes. These points are discussed in the literature on ordinary differential equations and to some extent elsewhere in this book. Almost all methods have been used by some authors and claimed to be superior for their problem in hand: it is generally a question of trading speed for robustness and in reservoir simulation as a whole there is no consensus.

Within the finite element approach there seems to be more agreement (there are fewer authors) and it accords with my own preferences. Transient flow problems with sizeable dispersion terms are often *stiff* in the sense of having a wide range of time scales. To achieve stability, explicit methods need to impose very strong restrictions on time step size and stiffly stable implicit schemes have found favour. For smooth problems or simulations tending to a steady state it is worth considering one of the self-stepping higher-order methods such as Gear's method;[25] for cases where wells or other boundary conditions are frequently being changed this may not be worth doing. For flow equations of hyperbolic type, that is without dispersion terms, explicit time stepping is generally favoured. Even then, methods which choose their order and time step size automatically can sometimes be used.[25]

In order to continue let us here adopt the implicit backward Euler method. The time derivative in eqn (25) is at time t^{n+1} replaced by a step from t^n to t^{n+1}

$$\frac{S_i^{n+1} - S_i^n}{t^{n+1} - t^n}$$

and, the important point, everywhere else in the equation S_i is taken as S_i^{n+1}. Equations (25) now become coupled nonlinear algebraic equations to be solved at each time step. If an explicit method were used, with all S_i evaluated at S_i^n except in the derivative, things would come out very much simpler.

Step 4

In practice the numerical solution of a system of nonlinear algebraic equations usually involves successive linearisation. There are several ways to do this but variants of the Newton–Raphson method seem to be the most popular. To illustrate this we rewrite eqns (25), with the time discretisation included, as a set of N equations

$$E_j(S_1, \ldots, S_N) = 0 \qquad j = 1, \ldots, N \tag{26}$$

for the unknown variables S_i. Note we have dropped the $n+1$ from S_i^{n+1}.

The standard Newton step is

$$S_i^{(k+1)} = S_i^{(k)} - J_{ij}^{-1} E_j(S_1^{(k)}, \ldots, S_N^{(k)}) \tag{27}$$

which is to be iterated to (sufficient) convergence. Note also that the superscript (k) refers to the number of Newton iterations and has nothing to do with the timestep. J_{ij} is the Jacobian matrix $\partial E_j/\partial S_i^{(k)}$ and should strictly be updated at each iteration. This expensive step can sometimes be avoided in simpler problems by a modified Newton method. At this stage in our procedure we have reduced the problem in iterating eqn (27) to successive solution of a set of $N \times N$ linear algebraic equations for $S_1^{(k)}, \ldots, S_N^{(k)}$.

Step 5
The solution of the system of linear equations, the fifth and last step, is crucial to the cost of the whole simulation. It lies in the innermost loop of the method and is used many times. In a large reservoir simulation typically 60 or 70 per cent of the computing time is spent on this step. Because the basis functions ϕ_j are localised near node j and zero away from it, variables labelled with i away from j tend not to occur in the jth equation. Thus a large number of the coefficients in the linear equations are zero—the system is sparse. Because the layout of elements can be arbitrary, one of the strengths of the finite element method is that these non-zero coefficients do not group into bands, as is usual in finite difference methods, but are scattered about in the equations.

This has led to some difficulty in the finite element method in finding efficient solvers which can exploit the sparsity. Many of the most efficient solvers used in conventional finite difference simulators make careful use of the regular structure of their equations and we cannot do the same with finite elements. There is, however, a very clever method, the frontal method invented by Irons, which is natural for finite elements and can be made very efficient. It is a direct, rather than iterative, method—essentially Gaussian elimination—which takes account of the sparsity automatically in the way it handles the elements. Its cost in computer operations depends on the number of nodes in the shortest direction across the reservoir—the frontwidth—in the manner

$$\text{Cost} \propto \text{number of nodal variables} \times (\text{frontwidth})^2$$

For two-dimensional problems this gives a fourth-power rise with size, but nonetheless an efficient coding of the frontal method[16,26] can compete successfully with the best alternative iterative methods on general problems. For three-dimensional problems this is no longer true. The operation count for direct methods, including the frontal method, is just too high. The lack, till now of efficient methods for 3-dimensional problems has in my opinion been one of the prime reasons for the slow growth in application of the finite element method to realistic simulations. In the last year or two matters have been improving. In particular it now appears as though the solvers based on the preconditioned conjugate gradient method can be usefully applied to finite element simulations with hopes of high efficiency.

4. CONNECTION WITH FINITE DIFFERENCE METHODS AND SOME EXAMPLES

For those readers unfamiliar with finite element methods the previous section will have seemed rather complicated. We wish in this section to show that the method can reasonably be thought of as a way of introducing high-order differencing on irregular meshes which in some cases reduces to finite difference formulae familiar to everyone. For this purpose the example in hand is too complicated and we simplify it again to the simple Poisson equation with sources. As this example has been much studied and documented we use a variation of it also as an illustration of some numerical results calculated by the finite element method.

The equation we use is

$$\nabla^2 P(x, y) = q(x, y) \tag{28}$$

on a two-dimensional square in x, y space. In a reservoir context this can be thought of as the pressure equation for a single-phase fluid in permeable rock. We take the simplest boundary conditions, that the normal derivative $\partial P/\partial n$ is zero on the boundary, and so need for consistency that the sources and sinks add up to zero, $\iint q(x, y)\,\mathrm{d}x\,\mathrm{d}y = 0$. We divide the square into N^2 little squares each $l \times l$ in size and take these as the finite elements. The vertices of the little squares are to be finite element nodes and we write ϕ_{ij} for the basis function associated with node i, j. Following the ideas described earlier the weak form of eqn (28) for the test function $h(x, y)$ is

$$-\iint \nabla P \,.\, \nabla h \,\mathrm{d}x\,\mathrm{d}y = \iint qh\,\mathrm{d}x\,\mathrm{d}y \tag{29}$$

The next step is to expand P in terms of the basis functions and nodal values $P_{I,J}$

$$P(x, y) = \sum_{I,J} P_{IJ}\phi_{IJ}(x, y) \qquad I, J = 1, \ldots, N \tag{30}$$

noting that *all* the nodes I, J are used since there are no boundary conditions which specify the value of P. [For a unique solution of eqn (28) a reference value of P ought to be specified somewhere, but we ignore this point for the moment.] We now form the algebraic discrete equations by taking for test function h each of the $\phi_{i,j}$ in turn

$$= \sum_{I,J} P_{IJ} \iint \nabla\phi_{IJ} \cdot \nabla\phi_{ij} \, \mathrm{d}x \, \mathrm{d}y = \iint q\phi_{ij} \, \mathrm{d}x \, \mathrm{d}y \qquad i,j = 1, \ldots, N \qquad (31)$$

The expression $\nabla f . \nabla g$ represents

$$\frac{\partial f}{\partial x}\frac{\partial g}{\partial x} + \frac{\partial f}{\partial y}\frac{\partial g}{\partial y}$$

Note that there is no time dependence in this problem and that because the original eqn (28) was linear the discrete eqns (31) are automatically linear. As ϕ_{ij} is a finite element basis function it is non-zero only on elements which contain the node i,j. As the integrals in eqns (31), which are over the large square, can be treated as the sum of integrals over the little squares it is easy to see that the left-hand side of eqns (31) is different from zero only for

$$I, J = \begin{array}{l} i,j;\ i-1,j;\ i+1,j;\ i,j-1;\ i,j+1; \\ i-1,j-1;\ i-1,j+1;\ i+1,j-1;\ i+1,j+1 \end{array}$$

so when written out explicitly the left-hand side is minus

$$\begin{aligned} & P_{ij} \iint \nabla\phi_{ij} \cdot \nabla\phi_{ij} + P_{i+1,j} \iint \nabla\phi_{i+1,j} \cdot \nabla\phi_{ij} + P_{i-1,j} \iint \nabla\phi_{i-1,j} \cdot \nabla\phi_{ij} \\ & \quad + P_{i,j-1} \iint \nabla\phi_{i,j-1} \cdot \nabla\phi_{ij} + P_{i,j+1} \iint \nabla\phi_{i,j+1} \cdot \nabla\phi_{ij} \\ & \quad + P_{i-1,j-1} \iint \nabla\phi_{i-1,j-1} \cdot \nabla\phi_{ij} + P_{i-1,j+1} \iint \nabla\phi_{i-1,j+1} \cdot \nabla\phi_{ij} \\ & \quad + P_{i+1,j-1} \iint \nabla\phi_{i+1,j-1} \cdot \nabla\phi_{ij} + P_{i+1,j+1} \iint \nabla\phi_{i+1,j+1} \cdot \nabla\phi_{ij} \end{aligned} \qquad (32)$$

Each of the integrals appearing in this expression is integrated over a small number of elements in which the overlap of the integrand factors is not zero. If we label as A, B, C, D the four elements around node *ij* in Fig. 4 then the integrals in eqn (32) are, in order, non-zero on the elements (ABCD), (BC), (AD), (AB), (DC), (A), (D), (B), (C). We now take on each small square the shape function of the bilinear Lagrangian or Serendipity element (they are the same[24]) in which the polynomial is $a + bx + cy + dxy$ with the four

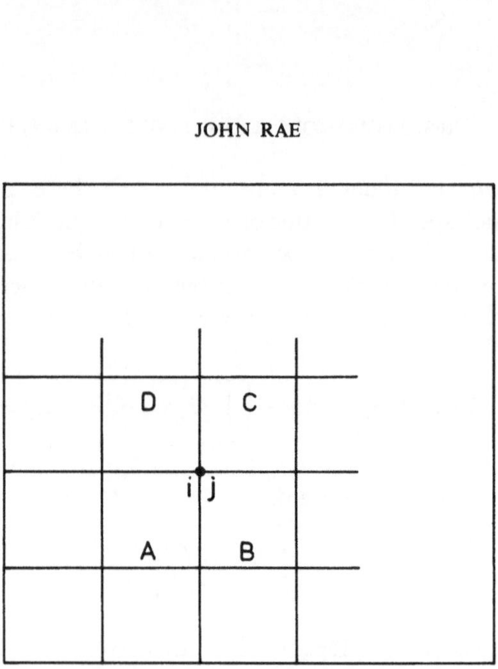

FIG. 4. Detail of the grid used to relate the finite element and finite difference methods.

coefficients fitted by the four nodal values at the vertices of the square. On the element C of Fig. 4, for example, the shape functions are

$$\begin{aligned} \phi_{ij} &= (x-(i+1)l)(y-(j+1)l)/l^2 \\ \phi_{i+1,j} &= -(x-il)(y-(j+1)l)/l^2 \\ \phi_{i,j+1} &= -(x-(i+1)l)(y-jl)/l^2 \\ \phi_{i+1,j+1} &= (x-il)(y-jl)/l^2 \end{aligned} \tag{33}$$

With these forms and a certain amount of symmetric transposition it is easy to work out the integrals in eqn (32). For example, the contribution from the element C to the first integral is

$$\begin{aligned} \iint_C \nabla\phi_{ij}\cdot\nabla\phi_{ij}\,\mathrm{d}x\,\mathrm{d}y &= \iint_C \left[\left(\frac{\partial\phi_{ij}}{\partial x}\right)^2 + \left(\frac{\partial\phi_{ij}}{\partial y}\right)^2\right]\mathrm{d}x\,\mathrm{d}y \\ &= \frac{1}{l^4}\int_{il}^{(i+1)l}\mathrm{d}x\int_{jl}^{(j+1)l}\mathrm{d}y\,[(x-(i+1)l)^2+(y-(j+1)l)^2] \\ &= \frac{2}{l^3}\int_{-l}^{0} t^2\,\mathrm{d}t \qquad \text{where } t = x-(i+1)l \\ &= \tfrac{2}{3} \end{aligned} \tag{34}$$

In a similar way we have

$$\iint_C \nabla\phi_{ij} . \nabla\phi_{i+1,j} \, dx \, dy = -\tfrac{1}{6} \quad (35)$$

$$\iint_C \nabla\phi_{ij} . \nabla\phi_{i+1,j+1} \, dx \, dy = -\tfrac{1}{3} \quad (36)$$

The integrals (eqns (34)–(36)) together with symmetric exchanges now let us write for eqn (32), the left-hand side of eqns (31)

$$\begin{aligned} &\tfrac{8}{3}P_{ij} - \tfrac{1}{3}(P_{i+1,j} + P_{i-1,j}P_{i,j-1} + P_{i,j+1}) \\ &\qquad -\tfrac{1}{3}(P_{i-1,j-1} + P_{i-1,j+1} + P_{i+1,j-1} + P_{i+1,j+1}) \end{aligned} \quad (37)$$

The expression (37) is just a representation of ∇^2. In fact if one expands eqn (37) in a Taylor series about the point i, j the result is

$$l^2\left(\frac{\partial^2 P}{\partial x^2} + \frac{\partial^2 P}{\partial y^2}\right) + 0(l^4) \quad (38)$$

Thus eqn (37) is just an order l^2 accurate finite difference representation of ∇^2 in terms of the point i, j and its eight neighbours and next-to-neighbours. It is not the most commonly used l^2 representation (see below) but is perfectly valid. In this simple case of a regular grid and low-order interpolation the advantages of the finite element method are not apparent; in fact it is more cumbersome than finite difference methods. But the relationship between the two should now be clear.

In the example above the integrals over shape functions (34)–(36) were worked out exactly by hand. When the equations are complex, the shape functions are of high order or curved elements are used, this is no longer possible and numerical integration is used. As this is only approximate (in general) it becomes, properly speaking, part of the discretisation approximation; the elements are accurate to some order in l (l^2 in eqn (37)) and a comparable accuracy would usually be used for the integration scheme. Gauss schemes are probably the most commonly used,[24] but there is one rather unusual choice which highlights the remarks just made.[27] The lowest-order Lobatto integration, the trapezoidal rule, gives the integral of any function $g(x, y)$ over the square C of Fig. 4 as one-quarter the sum of its values at the four corners times its area:

$$\iint_C g(x, y) \, dx \, dy \Big/ \frac{l^2}{4} [g_{ij} + g_{i+1,j} + g_{i,j+1} + g_{i+1,j+1}] \quad (39)$$

This integration formula has errors of order l^2 so it is not unreasonable to use it to evaluate integrals (34)–(36) rather than do them exactly. However, because of the way the shape functions are defined, they and their derivatives vanish at some of the four corners. For example $\partial\phi_{ij}/\partial x$ vanishes at $(i+1, j+1)$ and $(i, j+1)$ and takes the value $-1/l$ at (i, j) and $(i+1, j)$. When integral (34) is evaluated in the manner of eqn (39) it takes the value 1. Similarly the integrals (35), (36) come out with the values $-\frac{1}{2}$ and zero respectively. By making use of symmetries again the entire representation (eqn (32)) for ∇^2 can be re-evaluated. It does not now reduce to eqn (37) but rather to

$$4P_{ij} - P_{i+1,j} - P_{i-1,j} - P_{i,j-1} - P_{i,j+1} \tag{40}$$

which is perhaps the best known representation of ∇^2 in two-dimensions and is also order l^2 accurate. The appearance of eqn (37) and now eqn (40) emphasises that in the finite element method the integration scheme can be part of the discretisation, and also that conventional finite difference expressions are included as low-order regular element schemes.

As eqn (28) has been so much studied and should be reasonably familiar to readers, we continue to use it now to illustrate some actual computations with finite element methods. The example we will use is eqn (28) with one-point source and one-point sink in a circular reservoir, with P being regarded as the fluid pressure. Distances are scaled so that the reservoir has a radius of 1 unit and is centred at the x, y origin. Its curved boundary allows us to illustrate the use of curvilinear elements. We solve the problem as a two-dimensional x, y region and not as cylindrically symmetric. This lets us place the source and sink at $x = \pm 0{\cdot}25$, $y = 0$ and we give them strengths of -2 (injection into the reservoir) and $+2$ (production from the reservoir). With the source and sink of equal strength it is consistent to take no-flow boundary conditions on the edge of the reservoir.

The first series of results are presented for a relatively coarse grid of elements—the sort of grid one might try first. It is illustrated in Fig. 5 and divides the reservoir into 200 triangles, some with curved edges. No attempt has been made to refine the mesh near the source and sink which are represented by dots in the figure. The geometrical triangles can carry different shape functions and first we use linear triangles for which the vertices take independent nodal values and the interpolation is linear inside each triangle. This gives 120 nodal values to be calculated and one, on the middle of the top boundary which is fixed at the value zero to provide a reference pressure. Except when stated otherwise we use a seven-point Gauss integration on each triangle.

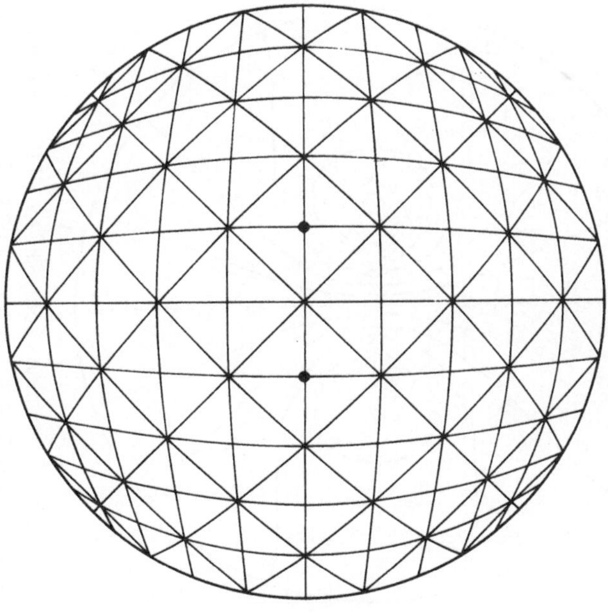

FIG. 5. The coarse grid for the computed example.

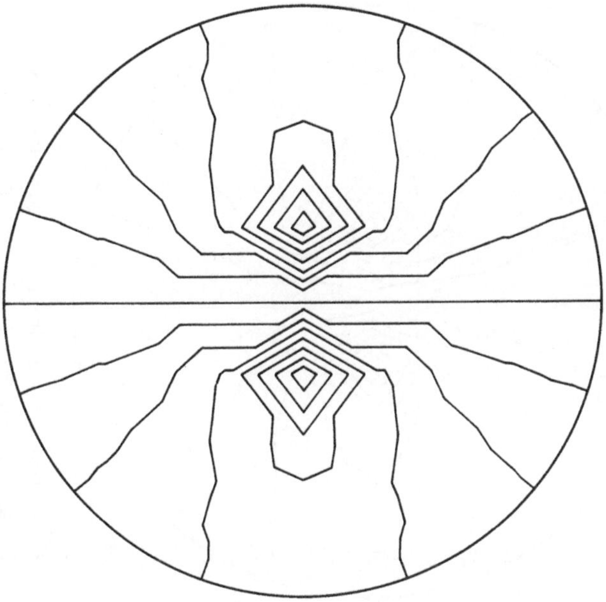

FIG. 6. Pressure contours from −0·7 to +0·7 in steps of 0·1; coarse grid, linear elements, seven-point Gauss integration.

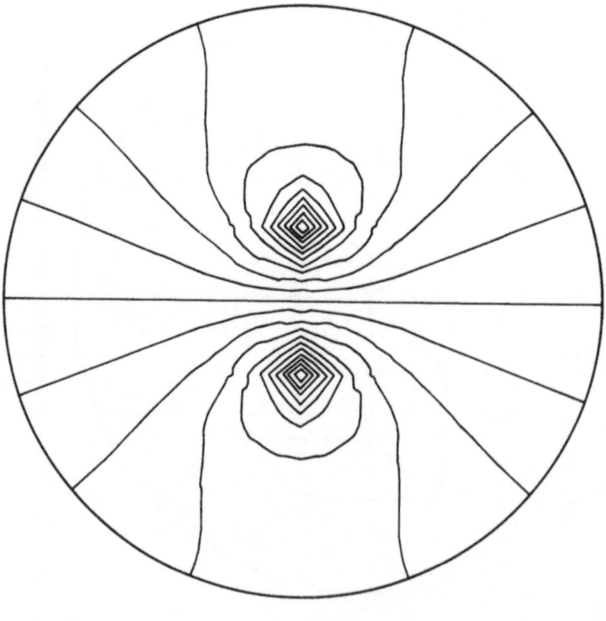

FIG. 7. Pressure contours from −1·0 to +1·0 in steps of 0·1; coarse grid, quadratic elements, seven-point Gauss integration.

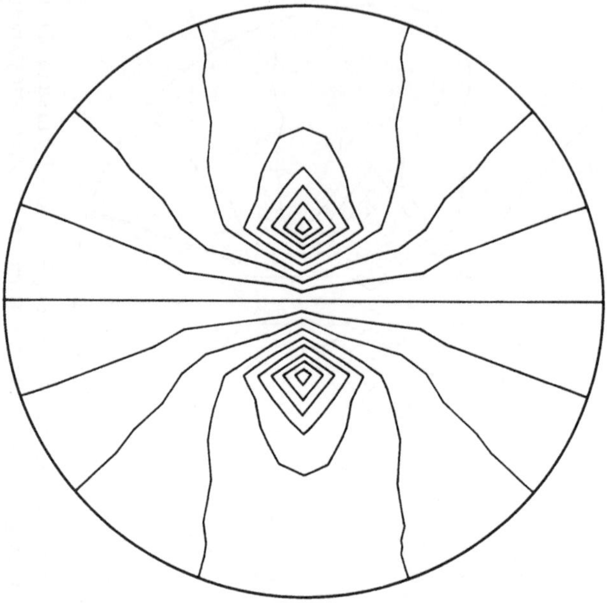

FIG. 8. Pressure contours from −1·0 to +1·0 in steps of 0·1; coarse grid, quadratic elements, three-point Gauss integration.

Figure 6 gives 15 pressure contours plotted from $-0{\cdot}7$ to $+0{\cdot}7$ in intervals of 0·1. The $-0{\cdot}7$ and $+0{\cdot}7$ values lie close to the sink and source respectively and the symmetry line is the contour with value zero. The results shown are very poor. The contours appear very jagged and clearly reflect the mesh structure in many places. The chosen mesh and linear shape functions are inadequate for the problem. Figure 7 shows what happens when we replace the linear shape functions by quadratics, keeping everything else the same. In Fig. 7 we are now solving for 440 nodal values and plot the 21 contours of pressure from $-1{\cdot}0$ to $+1{\cdot}0$ in steps of 0·1. These results are much improved but mesh effects are still evident, particularly near the source and sink. Next, we keep the same grid and quadratic shape functions, but change from the seven-point Gauss integration to one based on only three points, the vertices of the triangles. This Gauss scheme is very poor at sampling the quadratic functions and, as Fig. 8 shows, the results become worse again and we cannot resolve the contours near the source and sink. When a different three-point Gauss scheme was used, with the weight points inside the triangle, the sampling was so much better that the results are indistinguishable from those in Fig. 7.

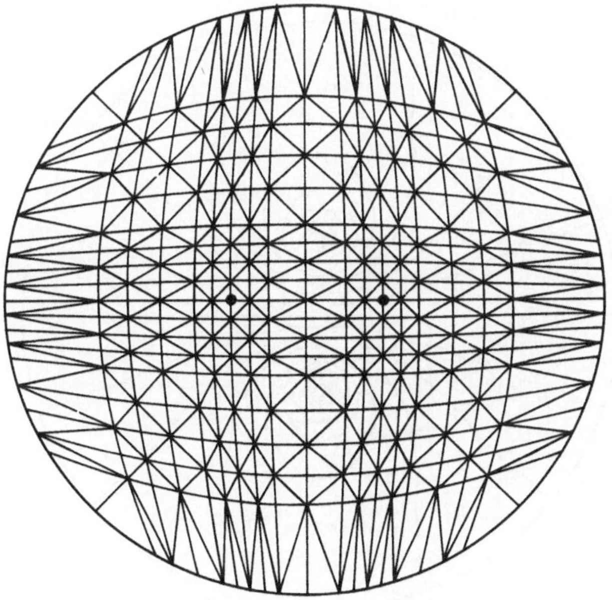

FIG. 9. The fine grid for the computed example.

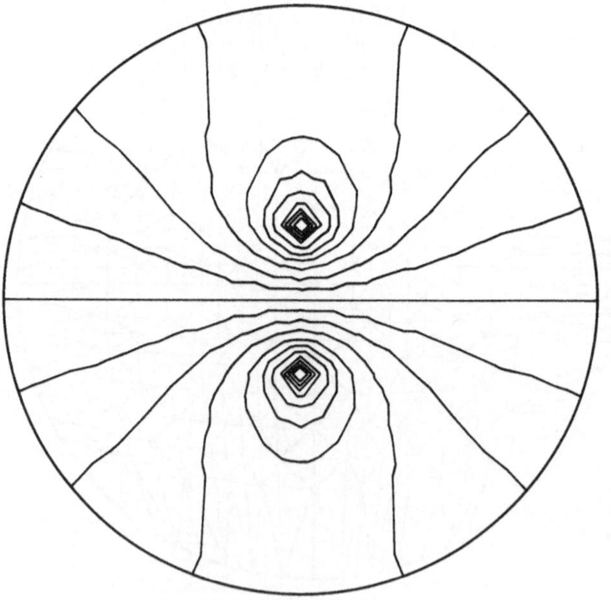

FIG. 10. Pressure contours from $-1{\cdot}0$ to $+1{\cdot}0$ in steps of $0{\cdot}1$; fine grid, linear elements, seven-point Gauss integration.

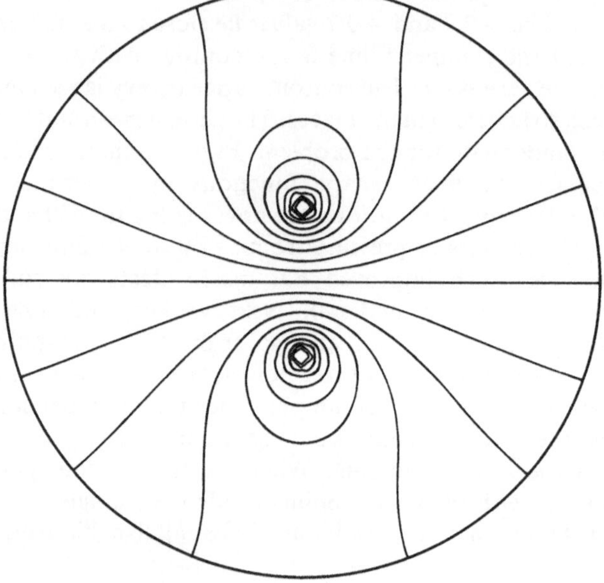

FIG. 11. Pressure contours from $-1{\cdot}0$ to $+1{\cdot}0$ in steps of $0{\cdot}1$; fine grid, quadratic elements, seven-point Gauss integration.

It is apparent that the mesh chosen in Fig. 5 is unable to resolve the spatial features of the problem and we must refine it. Figure 9 gives our second choice. There are now 648 triangles and we have graded their sizes so that the mesh is reasonably fine near the source and sink, indicated by dots, and coarse near the boundary where the pressure should be smoother. We then reran the problem with linear shape functions and the seven-point Gauss scheme. In Fig. 10 we plot 21 pressure contours from $-1{\cdot}0$ to $+1{\cdot}0$ in steps of 0·1. They are still a little jagged, especially at the source and sink, but are a clear improvement over the previous results. Finally, we change to quadratic shape functions on this same fine grid. The problem now has to be solved for 648 triangle and 1368 nodal pressure values. Pressure contours, again from $-1{\cdot}0$ to $+1{\cdot}0$ by steps of 0·1, are shown in Fig. 11. Near the source and sink the contours are a little square, but elsewhere they are quite acceptable. Pressure values at the source and sink should, of course, be infinite so there is in any case a limit to what we can achieve without special techniques such as singular shape functions. In Fig. 12 we give a pressure profile across the reservoir on the line through the wells. It is clear from this that the source and sink are not badly represented and for

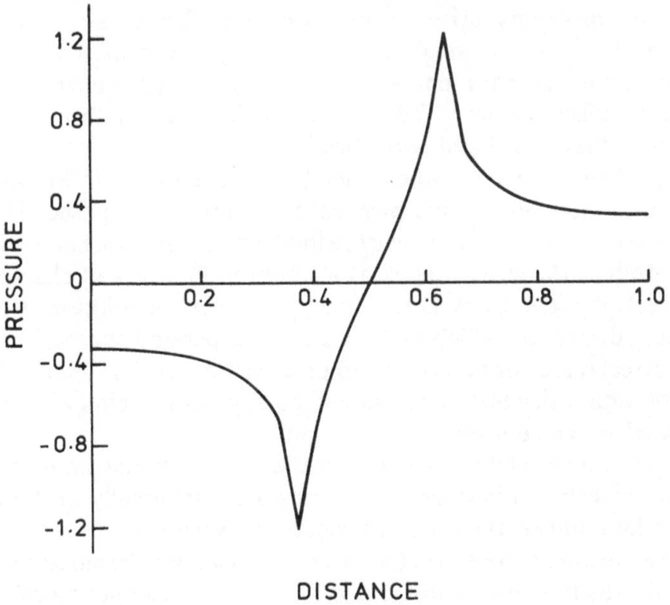

FIG. 12. Pressure profile across the reservoir on the line through the wells.

many practical purposes this last solution to the problem would be entirely acceptable. We therefore take the example no further.

Although it is sometimes risky to give computing times in isolation let me record that the program used for the example solved the equations by the frontal method mentioned earlier and took 6 seconds of cpu time on an IBM 3033 to find the 1368 pressure values after the mesh had been set up.

5. SPECIAL PROBLEMS

There are some special problems which arise when numerical methods are applied to reservoir simulations and, although it is impossible to go into detail in one short chapter, it is proper to mention briefly the impact of the finite element method on some of them.

One set of problems arises when dispersion effects are small, as when capillary pressure can be taken as constant. In this case the dispersion terms drop out of the governing equations, such as eqn (19), and they change their characters from parabolic to hyperbolic. Their solutions are no longer smooth but tend to develop discontinuities or sharp 'fronts' which can cause problems for numerical methods based on a fixed computational mesh. The problems arise in practice even for parabolic equations whenever dispersion is small enough that the transition zones of the solution, though smooth, change on a length scale smaller than the mesh spacing. In effect, we have to deal with shocks and therefore run into the usual difficulties associated with these.

One problem is that, mathematically, the equations no longer have a unique solution unless further physical conditions are imposed. This is not an academic point: it is all too easy to invent a numerical scheme which will in fact produce the wrong shock as a solution. One way of characterising the correct physical shock is to construct a series of solutions with ever-decreasing dispersion which pass through sharper and sharper transitions to the correct discontinuous limit. In practice this result is used in the form that a solution with a little dispersion of an appropriate kind will be close to the desired shock solution.

A second, and related problem is that conventional finite difference forms of advection–dispersion equations are numerically unstable if the mesh Peclet number (the ratio of time for dispersion across a mesh cell to time for advection across it) is too large. This can only be cured by refining the mesh, which is often ruled out on grounds of computing capacity or cost, or by increasing the dispersion in some way. The difficulty is, of course, to find a method of including just the right amount of dispersion to get round these problems. A number of rather sophisticated methods have been proposed (see for example Ref. 28) but for the most part these have only been applied to one-dimensional problems. The methods which are overwhelmingly used in practice are all some form of upstreaming[1-3] in which relative permeabilities and perhaps other properties are chosen to take some account of the direction of flow. This is equivalent to introducing extra dispersion into the equations and has the effect of capturing the right shock but also of spreading it. The top two diagrams of Fig. 13 illustrate this for a Buckley–Leverett type of displacement: the first diagram being the exact solution at some stage of production and the second being the dispersed front. The spreading is often unacceptably large, perhaps up to eight or ten mesh spacings and although many remedies have been proposed I think it fair to say that none is wholly successful.

The finite element method does not avoid these problems. As shown earlier, finite element formulations are usually equivalent to some high-order finite difference scheme and, as such, do not suffer so acutely from Peclet number limitations. With an element mesh comparable to that used for lower-order finite differences the critical value can be pushed up by factors from 2 to perhaps 10 although in special problems much grander claims have been made.[29] Extra help can be got at little extra cost if the troublesome regions of flow are known and the mesh is refined in that neighbourhood only. (It is not easy to do this for problems with moving

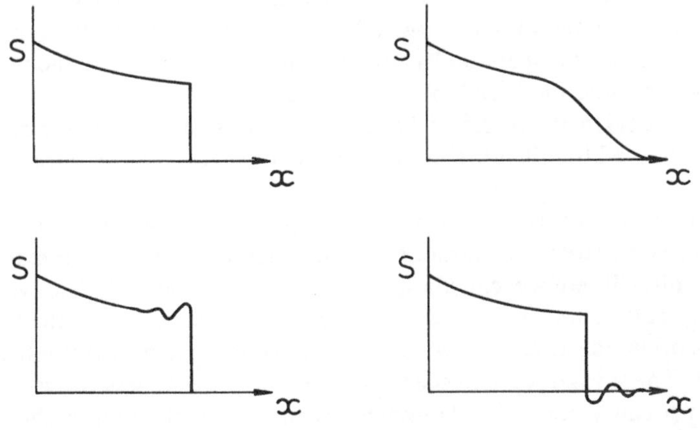

FIG. 13. Buckley–Leverett flood profiles illustrating spreading and oscillations of the shock.

fronts.) However, although the mesh stability problem is less serious for element meshes, other pitfalls await this approach. It is, for example, easy when there is no dispersion to compute the wrong shock solution, with a front which has too high a saturation value and moves too slowly.[30]

In addition, as shown in the bottom two diagrams of Fig. 13, ripples may appear in the solution before and behind the shock. All these effects can be understood from a truncation error analysis of the equations.[31] The incorrect speeds arise from the use of schemes without dispersion or diffusion. The ripples arise from two sources. The first is the use of high-order schemes and this applies whether elements or differences are being used. The second, the occurrence of capacity matrices, is specific to finite element formulations and needs a few lines of explanation. In eqn (25) the time derivative is multiplied by a matrix $\langle \phi_i, \phi_j \rangle$, known as the capacity or mass matrix, which is not diagonal. It is present even if explicit time differencing is used. In contrast, finite difference methods always lead to diagonal matrices in the time-derivative term. In explicit time stepping, then, no matrix inversion is needed for difference methods but the capacity matrix has to be inverted in element methods (but see Ref. 27). This is not really a problem; the matrix in our example has constant entries and is symmetric and it generally has reasonable structure and properties. However, to avoid the extra computing work it is fairly common practice to replace the capacity matrix by an effective diagonal or lumped matrix constructed from it. This increases the time truncation error and has undesirable side effects[35] but simplifies the computation. It is the use of unlumped capacity matrices which provokes the ripples. Many researchers in finite element methods are now following the lead of finite differences and introducing extra dispersion by up-streaming, often by introducing a bias into the test functions used in deriving the weak equations—the so-called Petrov–Galerkin method.[24] I have reservations about this for reservoir work; it has led to difficulties in other flow computations—see for example Ref. 32.

Another set of problems which crops up in the 'elements versus differences' debate is the question of grid orientation effects. In many low-order finite difference calculations, especially those with unfavourable mobility ratios, the computed results are rather sensitive to the way the mesh is oriented, something which is usually chosen arbitrarily and seldom revised. There is good evidence that the effect is much less for some finite element formulations.[29,30] However, recent results with finite differences, for example Ref. 33, show that the use of higher-order methods considerably reduces the effect. It appears, therefore, that element methods are superior in their handling of grid orientation effects but that this is a spin-off from their high-order character and ability to construct meshes in appropriately irregular ways.

6. CONCLUSIONS

Finite element methods have been applied in fluid mechanics problems only in the last 10 years and in reservoir simulation only much more recently. After some over-enthusiastic claims in the early days it has settled down to rather a fringe role in reservoir simulation with never quite enough users to take full benefit of its undoubted promise. With comparatively little experience to tap, it is rather too early to draw up any definitive list of merits, drawbacks, successes or failures, especially in comparison with the long-established finite difference method. Indeed, it may never be appropriate to do so, for both methods are still evolving and according to some recent trends even evolving towards each other. The following assessment should therefore be treated as tentative. The main advantages of the finite element method appear to be:

(i) It provides a natural way to implement high-order space discretisation with proper treatment of boundary conditions.
(ii) It allows a mix of different elements and local grid refinement in a way chosen almost freely by the user.
(iii) It can handle curved meshes without difficulty.
(iv) It allows elements with special, perhaps singular, shape functions to be included in the mesh at positions of known difficulty.
(v) Even on highly irregular meshes the method is on an excellent mathematical foundation for many types of equation.

The main disadvantages would appear to be:

(i) It is difficult to write a working program from scratch. Once a framework and set of service routines exist, as is now the case for example at Harwell, things become much easier, but such an environment is not at all common.
(ii) The high-order nature of the method often gives solutions which, although accurate, do display unphysical phenomena, for example an accurate representation of a sharp saturation peak marred by small wings of negative saturation. Many users object strongly.

(iii) Until now, linear equation solver techniques for finite element programs have been costly and generally inadequate for three-dimensional and very large two-dimensional problems. This has handicapped the application of the method. There are recent signs of progress here in adapting preconditioned conjugate gradient methods and the situation should improve.

The most interesting next steps for the finite element method look to be in its application to enhanced oil recovery processes. The accurate tracking of solute slugs or gas or steam floods might provide the sort of tough numerical problems in which its advantages can be exploited. Its record[34,35] in a large number of other fields of application makes it likely that it will grow in strength.

ACKNOWLEDGEMENTS

The author wishes to thank the United Kingdom Department of Energy, the British Gas Corporation and the British National Oil Corporation for support in this work and Miss E. A. Dolman for help in computing the figures.

REFERENCES

1. Crichlow, H. B. (1977) *Modern Reservoir Engineering, A Simulation Approach*, Prentice-Hall, Englewood Cliffs, New Jersey.
2. Aziz, K. and Settari, A. (1979) *Petroleum Reservoir Simulation*, Applied Science Publishers Ltd, London.
3. Peaceman, D. W. (1977) *Fundamentals of Numerical Reservoir Simulation*, Elsevier Scientific Publishing Co., Amsterdam.
4. Thomson, J. F. (1978) Numerical solution of flow problems using body-fitted coordinate systems. Lecture series No. 4, Von Karman Institute, Brussels.
5. Bischoff, K. B. and Levenspiel, O. (1962) Fluid dispersion—generalisation of models. *Chem. Eng. Sci.*, **17**, 245–54.
6. Williamson, A. S. and Chappelear, J. E. (1981) Representing wells in numerical reservoir simulation. *Soc. Pet. Eng. J.*, **21**(3), 323–38.
7. Zienkiewicz, O. C. and Heinrich, J. C. (1978) The finite element method and convection problems in fluid mechanics. In *Finite Elements in Fluids*, Vol. 3, Eds R. H. Gallagher *et al.*, Wiley, New York, pp. 1–22.
8. Yanosik, J. L. and McCracken, T. A. (1979) A nine-point finite-difference reservoir simulator for realistic prediction of unfavourable mobility ratio displacements. *Soc. Pet. Eng. J.*, **19**(4), 253–62.
9. Douglas, J., Darlow, B. L., Wheeler, M. and Kendall, R. P. (1979) Self-

adaptive Galerkin methods for one-dimensional two-phase immiscible flow. Paper 7679, *5th SPE Symposium on Reservoir Simulation*, Denver.
10. Gallagher, R. H., Zienkiewicz, O. C., Oden, J. T. and Taylor, C. (1976/78) *Finite Elements in Fluids*, Vols 1–3, Wiley, New York.
11. Connor, J. J. and Brebbia, C. A. (1976) *Finite Element Techniques for Fluid Flow*, Newnes-Butterworth, London.
12. Chung, T. J. (1978) *Finite Element Analysis in Fluid Dynamics*, McGraw-Hill, New York.
13. Szabo, B. A. (1977) Self-adaptive finite element approximations. *Proceedings of the 14th Annual Meeting of the Society of Engineering Science*, Lehigh University, Bethlehem, Penn.
14. Irons, B. M. (1970) A frontal solution program. *Int. J. Num. Meth. Eng.*, **2**, 5–32.
15. Hood, P. (1976) Frontal solution program for unsymmetric matrices. *Int. J. Num. Meth. Eng.*, **10**, 379–99.
16. Duff, I. S. (1981) MA32, A package for solving sparse unsymmetric systems using the frontal method. Harwell Report AERE R.10079.
17. Strang, G. and Fix, G. J. (1973) *An Analysis of the Finite Element Method*, Prentice-Hall, Englewood Cliffs, New Jersey.
18. Norrie, D. and DeVries, G. (1976) *A Finite Element Bibliography*, Plenum Press, New York.
19. Finlayson, B. A. (1972) *The Method of Weighted Residuals and Variational Principles*, Academic Press, New York.
20. Gradshteyn, I. S. and Ryzhik, I. W. (1965) *Table of Integrals, Series and Products*, Academic Press, London.
21. Courant, R. (1943) Variational methods for the solution of problems of equilibrium and vibration. *Bull. Am. Math. Soc.*, **49**, 1–23.
22. Oden, J. T. and Reddy, J. N. (1976) *An Introduction to the Mathematical Theory of Finite Elements*, Wiley, New York.
23. Douglas, J. and Dupont, T. (1970) Galerkin methods for parabolic equations. *Siam J. Numer. Anal.*, **7**, 575–626.
24. Zienkiewicz, O. C. (1977) *The Finite Element Method*, McGraw-Hill, London.
25. Gear, C. W. (1977) Ordinary differential equation techniques for partial differential equations. In *Formulations and Computational Algorithms in Finite Element Analysis*, Eds K. J. Bathe, J. T. Oden and W. Wunderlich, MIT, Cambridge, Mass.
26. Reid, J. K. (1981) Frontal methods for solving finite element systems of linear equations. In *Sparse Matrices and Their Uses*, Academic Press, London, pp. 265–81.
27. Young, L. C. (1981) An efficient finite element method for reservoir simulation, *Soc. Pet. Eng. J.*, **21**(1), 115–28.
28. Roe, P. L. and Baines, M. J. (1981) Algorithms for advection and shock problems. *4th GAMM Conference on Numerical Methods in Fluid Mechanics*, Paris.
29. Settari, A., Price, H. S. and Dupont, T. (1977) Development and application of variational methods for simulation of miscible displacement in porous media. *Soc. Pet. Eng. J.*, **17**(3), 228–46.

30. Spivak, A., Price, H. S. and Settari, A. (1977) Solution of the equations for multidimensional, two-phase, immiscible flow by variational methods. *Soc. Pet. Eng. J.*, **17**(1), 27–41.
31. Jackson, C. P. and Rae, J. (1979) The effects of truncation error in numerical modelling of flood fronts in oil reservoirs. Harwell Report TP.813.
32. Gresho, P. M. and Lee, R. L. (1981) Don't suppress the wiggles—they're telling you something. *Computers and Fluids*, **9**, 223–53.
33. Abou-Kassem, J. and Aziz, K. (1982) Grid orientation during steam displacement. Paper 10497, *6th SPE Symposium on Reservoir Simulation*, New Orleans, Louisiana; *Soc. Pet. Eng. J.* (1984), **24**(1), 65–74.
34. Hughes, T. J. R. (1980) Some current trends in finite element research. *App. Mech. Rev.*, **33**(11), 1467–77.
35. Gresho, P. M., Lee, R. L. and Sani, R. L. (1978) Advection-dominated flows with emphasis on the consequences of mass lumping. In *Finite Elements in Fluids*, Vol. 3, Eds R. H. Gallagher *et al.*, Wiley, New York, pp. 335–50.

Chapter 7

RESERVOIR MODELLING: FULLY IMPLICIT SIMULATION METHODS

J. R. APPLEYARD and I. M. CHESHIRE
Exploration Consultants Ltd, Henley-on-Thames, Oxon, UK

1. INTRODUCTION

Traditionally, the use of fully implicit methods in reservoir simulation has been reserved for special applications such as single well coning studies. However, recent advances in numerical methods have made their use attractive for larger problems, including full-scale three-dimensional field studies.[1–3] In this chapter, we examine the advantages of fully implicit methods and review the advances which have made their application to field-scale problems practicable.

Fully implicit simulations are characterised by the fact that flow rates, both within the reservoir and through wells, are computed using the solution (i.e. phase pressures and saturations throughout the reservoir) at the end of each time step. The term 'implicit' is used because of the interdependence of the flow rate during a time step and the solution at the end of it. The evaluation of consistent flow rates and solutions involves the solution of non-linear equations using an iterative technique such as the Newton–Raphson method. Other, less implicit methods involve less work at each time step, but are, as a result, prone to instability. The fully implicit method is stable for all time step lengths, and guarantees that saturations remain within physical bounds. These guarantees are of great practical importance in programs designed for use by engineers with limited time to devote to any particular study. In particular, they make the program much more robust, and eliminate the need to trace and correct numerical problems.

The unconditional stability of the fully implicit method makes it possible for the simulator to take much longer time steps than would be possible for a more explicit program. Increasing time step lengths reduces the total computer time required to complete a given simulation, with the result that fully implicit methods are often the most efficient as well as the most robust. However, longer time steps also result in larger time truncation errors, which often result in smearing of saturation fronts and premature breakthrough at wells. Whilst the effect of time truncation errors should not be overestimated, methods which reduce them are of considerable interest. A new method which achieves this goal without sacrificing the unconditional stability of the fully implicit method has recently been introduced[4] and is reviewed in Section 3 of this chapter.

As mentioned above, use of the fully implicit method involves the solution, at each time step, of sets of coupled non-linear equations. In general, Newton's method is used to solve these equations, and it is found that the difficulty in obtaining a solution increases with the time step length. In Section 4 we review these problems, and describe a new technique (the 'cascade' method) for improving the rate of convergence. This technique has not, to the best of our knowledge, been described elsewhere.

The object of the cascade method is to reduce the number of Newton iterations required to solve the non-linear equations. The speed of the simulator could also be increased by reducing the amount of work done at each Newton iteration, provided that this does not degrade the convergence rate. Often, this is done by decoupling the phases so that the equations can be solved sequentially (by solving for pressures first and subsequently for saturations). In Section 5, we review some procedures which have been used to achieve this decoupling. Currently, the most promising appears to be the 'adaptive implicit' method of Thomas and Thurnau.[5,6] However, it is doubtful whether the speed improvements obtained using the adaptive implicit method with a direct linear equation solver can also be achieved with an iterative solver. Until this is demonstrated, the applicability of the method to field-scale simulations remains in doubt.

In recent years, very significant advances have been made in techniques for iterative solution of the linearised equations at each Newtonian iteration. To date, the most successful methods have been based on the preconditioned conjugate gradient algorithm of Meijerink and Van der Vorst.[7] These methods are reviewed in Section 5 of this chapter. First, the conjugate gradient method is described using a formulation which emphasises the relationship between the original algorithm (which is valid only when the coefficient matrix is symmetric and positive definite) and the orthomin procedure of Vinsome[8] which is applicable to the asymmetric problems which arise in reservoir simulation.

The most important factor affecting the rate of convergence of these methods is the choice of preconditioning matrix. Most workers have used an incomplete Cholesky factorisation of the coefficient matrix, and this type of preconditioning is discussed in Section 5. Recently, a new preconditioning method which is not based on Cholesky factorisation has been introduced by Appleyard *et al.*[4] This procedure, known as nested factorisation, appears to be superior to incomplete Cholesky factorisation in both computer time and storage requirements. Because the method is not yet widely known we describe it in some detail.

2. THE FINITE DIFFERENCE EQUATIONS

This chapter is devoted to a description of the finite difference equations used in black oil and pseudo-gas condensate reservoir simulation. We show that the equations can be summarised using a very simple notation which we will use throughout the remainder of the chapter. The notation is not restricted to the special case of black oil simulation, and can be used for other types of problem.

As a starting point we shall assume that the equations have been discretised in space but not in time. The space discretisation procedure has been described thoroughly in the literature and the reader is referred to standard texts on the subject for further details.[9]

The equations describing the flow of fluids in oil reservoirs can be summarised in the form

$$\frac{\mathrm{d}M}{\mathrm{d}t} = F + Q \tag{1}$$

where M is the mass of a given component in a cell, F is the rate at which the component is flowing into the cell from neighbouring cells, and Q is the rate of production or injection through wells (Q is negative for a production well).

For any given cell in the reservoir, there is one such equation for each component. M, F and Q can be regarded as vectors with one vector element for every cell. Each vector element contains one number for each fluid component. In a three-phase black oil simulator, elements of M have the form

$$M_i = V_i \begin{pmatrix} \dfrac{S_o}{B_o} + R_v \dfrac{S_g}{B_g} \\ \dfrac{S_w}{B_w} \\ \dfrac{S_g}{B_g} + R_s \dfrac{S_o}{B_o} \end{pmatrix}_i \tag{2}$$

where: V_i is the pore volume of cell i; S_o, S_w and S_g are the saturations of oil, water and gas, respectively; B_o, B_w and B_g are the formation volume factors of oil, water and gas, respectively; R_s is the ratio of solution gas to oil; R_v is the ratio of vapourised oil to free gas.

Elements of F are formed as the sum of flows from all neighbouring cells.

$$F_i = \sum_n F_{ni} \tag{3}$$

The rate of flow from a neighbouring cell n into cell i, F_{ni}, is given by

$$F_{ni} = T_{ni} \begin{pmatrix} \dfrac{k_{ro}}{B_o\mu_o} & 0 & R_v \dfrac{k_{rg}}{B_g\mu_g} \\ 0 & \dfrac{k_{rw}}{B_w\mu_w} & 0 \\ R_s \dfrac{k_{ro}}{B_o\mu_o} & 0 & \dfrac{k_{rg}}{B_g\mu_g} \end{pmatrix}_u \begin{pmatrix} \Delta\Phi_{oni} \\ \Delta\Phi_{wni} \\ \Delta\Phi_{gni} \end{pmatrix} \tag{4}$$

where the subscripts o, w and g refer to oil, water and gas, and

$$\begin{aligned} \Delta\Phi_{oni} &= P_{on} - P_{oi} - \rho_{oni} g(d_n - d_i) \\ \Delta\Phi_{wni} &= P_{on} - P_{oi} - \rho_{wni} g(d_n - d_i) - (P_{cown} - P_{cowi}) \\ \Delta\Phi_{gni} &= P_{on} - P_{oi} - \rho_{gni} g(d_n - d_i) + (P_{cogn} - P_{cogi}) \end{aligned} \tag{5}$$

Here, k_r is the relative permeability, μ is the viscosity, P is the phase pressure, ρ is the fluid density, g is the acceleration due to gravity, d is the cell depth, P_{cow} is the capillary pressure difference in the oil and water phases, P_{cog} is the capillary pressure difference in the oil and gas phases and T_{ni} is the transmissibility between cells n and i (a function of the grid geometry and the absolute permeability of the reservoir).

The subscript u in eqn (4) indicates that fluid mobilities are to be computed in the upstream cell (cell n if $\Delta\Phi_{ni}$ is positive, or cell i if $\Delta\Phi_{ni}$ is negative).

The rate of flow into a production well from cell i is given by

$$Q_i = -T_{iw}(P_{oi} - H_{iw} - P_{bh}) \begin{pmatrix} \dfrac{k_{ro}}{B_o\mu_o} + R_v \dfrac{k_{rg}}{B_g\mu_g} \\ \dfrac{k_{rw}}{B_w\mu_w} \\ \dfrac{k_{rg}}{B_g\mu_g} + R_s \dfrac{k_{ro}}{B_o\mu_o} \end{pmatrix}_i \tag{6}$$

where P_{bh} is the well flowing pressure at the bottom hole, H_{iw} is a hydrostatic head correction and T_{iw} is, effectively, a transmissibility between the cell i and the wellbore.

For a well operating at a fixed flow rate Q^c of the controlled phase, the bottom hole pressure, P_{bh}, is obtained from the condition

$$\sum_i Q_i = Q^c \tag{7}$$

with the summation taken over all cells in which the well is completed.

Finally, we note that the solution of the coupled eqns (1) is a state vector, X, which for cell i can be written as a three-component vector

$$X_i = \begin{pmatrix} P_o \\ S_w \\ S_g \end{pmatrix}_i \tag{8}$$

The remaining phase pressures and saturations are determined by the relations

$$\begin{aligned} P_w &= P_o - P_{cow}(S_w) \\ P_g &= P_o + P_{cog}(S_g) \\ S_o &= 1 - S_w - S_g \end{aligned} \tag{9}$$

Throughout the remainder of this article we shall use the dot notation for derivatives with respect to time

$$\dot{M} = \frac{\mathrm{d}M}{\mathrm{d}t} \tag{10}$$

and the prime for derivatives with respect to the state vector

$$F' = \frac{\partial F}{\partial X} \tag{11}$$

Note that $\dot{M}$ is a vector while F' is a matrix

$$F'_{ij} = \frac{\partial F_i}{\partial X_j} = \begin{pmatrix} \frac{\partial F_o}{\partial P} & \frac{\partial F_o}{\partial S_w} & \frac{\partial F_o}{\partial S_g} \\ \frac{\partial F_w}{\partial P} & \frac{\partial F_w}{\partial S_w} & \frac{\partial F_w}{\partial S_g} \\ \frac{\partial F_g}{\partial P} & \frac{\partial F_g}{\partial S_w} & \frac{\partial F_g}{\partial S_g} \end{pmatrix}_{ij} \tag{12}$$

3. TIME DISCRETISATION

By contrast with explicit and semi-implicit methods the fully implicit method is unconditionally stable and always gives physically meaningful results. However, significant time truncation errors are obtained if very large time steps are used. A modification of the fully implicit approximation which minimises time truncation errors is discussed.

3.1. The Fully Implicit Method

At each time step in a simulation the solution is advanced by integrating the mass conservation equations. Thus if at any instant we have

$$\dot{M} = F + Q \tag{13}$$

then for the interval from t to $t + \Delta t$, we have

$$\Delta M = M(t + \Delta t) - M(t) = \int_t^{t+\Delta t} (F(\tau) + Q(\tau))\,\mathrm{d}\tau \tag{14}$$

In most practical situations, the integral in eqn (14) cannot be performed analytically, and some numerical approximation must be made. The choice of approximation has a profound effect on the stability and accuracy of the simulation. In this section, we review some possible methods for approximating the integral in eqn (14), and compare their stability and accuracy.

Most approximations depend on assumptions about the way the solution variables (pressures and saturations) vary in time. The simplest of these is the fully explicit approximation, in which it is assumed that pressures and saturations remain constant at their initial values throughout each time step. With this assumption, eqn (14) can be written as

$$\Delta M = [F(P^t, S^t) + Q(P^t, S^t)]\,\Delta t \tag{15}$$

Note that in this case, the interblock and well flows for the time step are computed without knowledge of the way the solution variables are changing. As a result, methods based on the fully explicit approximation require very little work, but are very prone to instability. For example, the procedure may predict that the total mass of a fluid within a cell becomes negative. Indeed, for sufficiently long time steps, unphysical predictions of this sort are inevitable. In reservoir simulation, these considerations limit the time step to such a small value that the fully explicit method is of no practical use.

In the IMPES (IMplicit Pressure Explicit Saturation) approximation, saturations are, as before, assumed to remain constant at their initial values throughout the time step. Pressures, on the other hand, are assumed to reach their final values very quickly and to remain there throughout the time step,

$$\Delta M = [F(P^{t+\Delta t}, S^t) + Q(P^{t+\Delta t}, S^t)]\,\Delta t \tag{16}$$

In this case, the flow rates and the pressures at the end of the time step are interdependent and must be determined by iteration. This increases the work required to advance the simulation through a time step, but reduces the chance of instability.

The increased stability results from the fact that flow rates respond to changes in the pressure field in such a way as to damp excessive changes in the solution variables. It is, however, still possible for instabilities to arise as a result of the explicit treatment of saturations. The resulting limit on time step sizes is much less severe than for the fully explicit method, but is, nevertheless significant.

An unconditionally stable method is obtained by treating both pressures and saturations implicitly (the fully implicit approximation):

$$\Delta M = [F(P^{t+\Delta t}, S^{t+\Delta t}) + Q(P^{t+\Delta t}, S^{t+\Delta t})]\,\Delta t \tag{17}$$

This scheme cannot give rise to non-physical saturations. Saturations cannot fall below zero because a fluid can flow only if it is mobile at the end of a time step, and fluids are mobile only for saturations greater than zero. Saturations cannot exceed unity because this would imply negative saturations in other phases.

Traditionally, full-field reservoir simulations have been conducted using IMPES time discretisation, whilst fully implicit methods have been reserved for particularly testing applications, such as single well coning studies. In recent years, however, there has been a trend towards the use of fully implicit methods for other applications including full field studies. This shift has been motivated principally by the greater robustness of fully implicit methods. Not only are unphysical saturations excluded, but also such indications of instability or near-instability as oscillations in produced gas/oil ratio or water are cut.

As a result, the engineer can have much greater confidence that the results he obtains are a consequence of the data input, rather than some mathematical quirk of the simulator. In addition, the computing cost of fully implicit methods has been greatly reduced by recent advances in numerical methods, particularly in the solution of linear equations. In consequence, fully implicit methods are often the most efficient as well as the most robust alternative.

3.2. Time Truncation Errors

The errors which result from using one of the approximations described above rather than performing the integral in eqn (14) exactly are known as time truncation errors. Figure 1 shows how these errors arise.

The object of all the approximations is to estimate the total mass accumulation, which is given by the area under the curve $(A+B+C)$. The explicit approximation gives the lower rectangle (area A), whilst the implicit approximation gives the larger rectangle (area $A+B+C+D$).

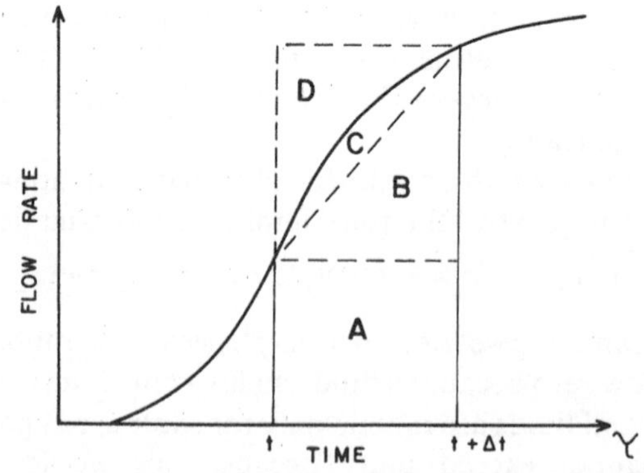

FIG. 1. Interblock flow rate versus time for a small time step, Δt.

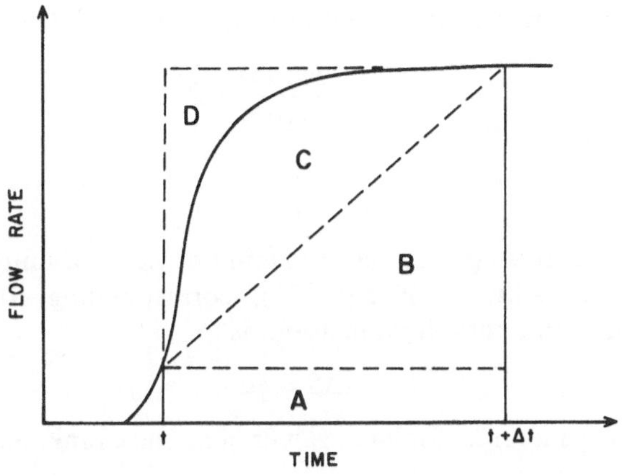

FIG. 2. Interblock flow rate versus time for a large time step, Δt.

Apparently the most accurate approximation is generated by the time centred (Crank–Nicholson) method which uses the average of the implicit and explicit values (area $A + B$). For short time steps, over which the flow varies linearly with time, the Crank–Nicholson method is indeed the most accurate. However, for larger time steps, there is typically a transient period after which the flow settles to a constant value, and in such cases, the fully implicit approximation is likely to be most accurate (see Fig. 2).

3.3. An Illustrative Example

To help clarify the ideas introduced above we shall now discuss a simple numerical example. Consider a tank with water flowing in at a constant rate q and flowing out at a rate qS^2 where S is the saturation or fraction of tank already filled. The mass of water in the tank is

$$M = S\rho \tag{18}$$

and the net inflow rate is

$$F = \rho q(1 - S^2) \tag{19}$$

where ρ is the density of water. The conservation equation governing the contents of the tank is

$$\dot{S} = q(1 - S^2) \tag{20}$$

and assuming that the tank is initially empty the exact solution is

$$\Delta S = \frac{\exp(\alpha) - 1}{\exp(\alpha) + 1} \tag{21}$$

where

$$\alpha = 2qt \tag{22}$$

and for large time steps, ΔS has a maximum value of unity.

The explicit solution of eqn (20), corresponding to the IMPES approximation in reservoir simulation, is

$$\Delta S = \tfrac{1}{2}\alpha \tag{23}$$

which is clearly incorrect for large values of α. The Crank–Nicholson result is

$$\Delta S = \frac{\alpha}{1 + \left(1 + \dfrac{\alpha^2}{2}\right)^{1/2}} \tag{24}$$

which approaches $\sqrt{2}$ for large values of α. Finally the fully implicit approximation gives

$$\Delta S = \frac{\alpha}{1 + (1 + \alpha^2)^{1/2}} \tag{25}$$

which has the correct asymptotic form for large time steps.

3.4. Reduction of Time Truncation Errors

Recently, Appleyard *et al.*[4] have proposed a new time discretisation scheme which aims to minimise time truncation errors while retaining the unconditional stability of the fully implicit method.

Their model treats pressures implicitly, but performs the integral over saturation analytically using a simple model for the variation of saturation with time.

$$\begin{aligned} \Delta M &= \int_t^{t+\Delta t} [F(P^{t+\Delta t}, S^\tau) + Q(P^{t+\Delta t}, S^\tau)]\,\mathrm{d}\tau \\ &= \int_{S^t}^{S^{t+\Delta t}} [F(P^{t+\Delta t}, S) + Q(P^{t+\Delta t}, S)]\,\frac{\mathrm{d}S}{\dot{S}} \end{aligned} \tag{26}$$

where

$$\dot{S} = \theta(\delta - f(S)) \tag{27}$$

and θ is an arbitrary constant, f is the fractional flow for the current phase, $\delta = 1$ for an invading phase (so that $\dot{S} \to 0$ as $f \to 1$) or $\delta = 0$ for a displaced phase (so that $\dot{S} \to 0$ as $f \to 0$).

In the absence of capillary pressure, the method corresponds to replacing the relative permeabilities, $k_r(S + \Delta S)$, in eqn (4) by their time averaged values $\bar{k}_r(S, S + \Delta S)$ given by

$$\bar{k}_r \int_{S^t}^{S^{t+\Delta t}} \frac{dS}{\delta - f} = \int_{S^t}^{S^{t+\Delta t}} \frac{k_r \, dS}{\delta - f} \tag{28}$$

In simple cases, for example where relative permeability is a piecewise linear function of saturation, the integral in eqn (28) can be performed analytically for every cell at minimal computing cost. Unlike other approximations to the flow integral, this method takes full account of the shape of the relative permeability curve throughout the interval S to $S + \Delta S$, and not just values at the end points.

It may be shown that, in common with the exact solution, the above scheme tends towards a time centred (Crank–Nicholson) approximation for small ΔS, and towards the fully implicit value for large ΔS. Moreover, the fact that $\dot{S}$ tends to zero as the saturation approaches a physical bound ensures that, as with the fully implicit method, unphysical saturations cannot occur.

The results presented in Ref. 4 support the view that the above method is as stable as a fully implicit scheme, but much more accurate (the number of time steps required to achieve a given accuracy is typically reduced by an order of magnitude). However, the method as described can be applied only to two-phase problems. Moreover, it is not clear that the use of a simple form of fractional flow can be justified in cases where fluids are flowing in opposite directions (as in gravity segregation). Some further research is therefore required to generalise the method for practical use.

4. SOLUTION OF THE NON-LINEAR EQUATIONS

At each time step, the fully implicit method gives rise to non-linear equations which are usually solved by repeated linearisation using Newton's method. The iteration may fail to converge if the problem is highly non-linear, or when very large time steps are taken. A new method is described to accelerate Newton's method on difficult problems.

4.1. Newton's Method

The fully implicit equations which must be solved at each time step are of the form

$$R = \frac{\Delta M}{\Delta t} - F(t + \Delta t) - Q(t + \Delta t) = 0 \tag{29}$$

where the mass accumulation, ΔM, the interblock flow, F, and the well flow, Q, are all functions of the state vector, X, evaluated at the end of the time step (time $t + \Delta t$). These equations are normally solved by Newton's method. Thus, if X_n is the current best estimate of the solution, and $R(X_n)$ is the corresponding residual, then to first order,

$$R(X_n + \delta X) = R(X_n) + R'(X_n)\,\delta X \tag{30}$$

and the iteration is defined by

$$X_{n+1} = X_n - [R'(X_n)]^{-1} R(X_n) \tag{31}$$

which is a generalisation to many variables of the single-variable procedure illustrated in Fig. 3. For a starting point which is sufficiently close to the solution, Newton's method gives quadratic convergence (i.e. the number of correct digits in the solution doubles at each iteration). Great care must be

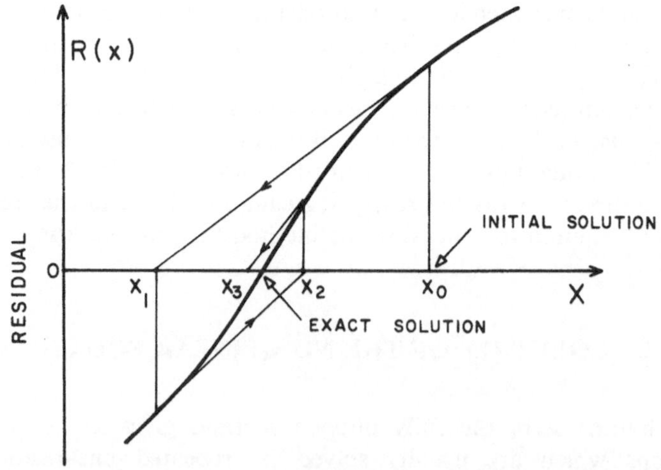

FIG. 3. Illustration of Newton's method for a one-dimensional problem. Starting from an initial approximation, X_0, the method converges through intermediate solutions, X_1, X_2, X_3, etc., to the exact solution.

taken to ensure that the Jacobian matrix, R', is properly evaluated, since even apparently minor omissions will cause the convergence rate to be linear, at best.[10] Indeed, the quadratic convergence property provides a valuable check to help ensure that the simulator has been correctly programmed.

However, Newton's method may not converge at all if the starting point is far from the solution. In Fig. 4 for example, the Newton iteration oscillates between X_1 and X_2. Such behaviour is possible wherever R'' changes sign (i.e. for S-shaped curves). It is quite simple to devise a procedure which ensures convergence in such cases for single-variable problems—for example, by ensuring that the solution is inside a contracting boundary. However, such methods cannot easily be applied to many-variable problems. In reservoir simulation, convergence difficulties are often caused by non-linearities in the relative permeability curves. In particular, the transition between mobility and immobility is often difficult.[11] Problems may also arise when using pseudo-relative permeabilities, which fall to zero below the Buckley–Leverett saturation, to control numerical dispersion.[4]

In the next section we describe a new technique which aims to eliminate these difficulties by reducing the many-variable problem to a large number of single-variable problems.

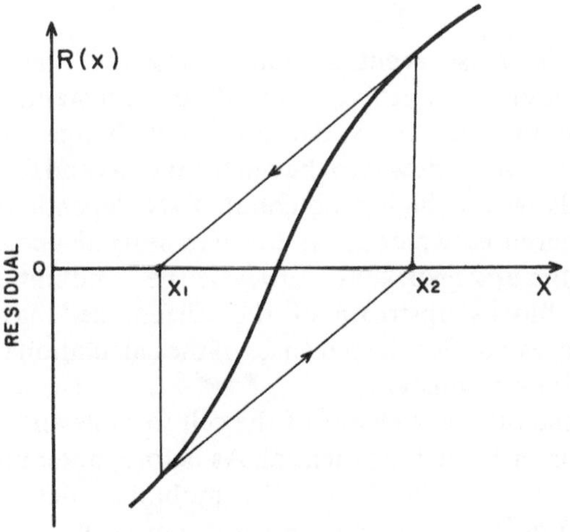

FIG. 4. Failure of Newton's method. Here, the technique oscillates between X_1 and X_2.

4.2. The Cascade Method

The cascade method is a procedure for updating the phase saturations (but not the phase pressures) at each Newton iteration using a direct method to solve the non-linear equations. Each iteration has two steps:

1. The cascade step
   Assuming that phase pressures remain unchanged, obtain the saturation of each phase in every cell by direct solution of the non-linear equations.
2. The Newton correction
   If the saturations sum to unity in every cell of the reservoir, a consistent set of phase saturations and pressures has been found, and the equations are solved. If not, update the pressures using Newton's method. In general, it is necessary to solve for all phases simultaneously.

In step 1, the saturation updates are computed separately for each phase. The calculation starts from the cell with the highest potential and works in potential order to the cell with the lowest potential. It will be noted that the cells will be ordered differently for different phases. Consider a cell at any point in the sequence. The mass conservation equation for this cell takes the form

$$\frac{\Delta M(S)}{\Delta t} = \mathrm{FIN} - \mathrm{FOUT}(S) \tag{32}$$

where ΔM is the mass accumulation during the time step. As phase pressures are to be considered constant, ΔM is simply proportional to ΔS, the saturation change.

FIN is the rate of flow into the cell from wells and from neighbouring cells with a higher potential. FIN depends on the potential differences, which are assumed to be fixed, and on the mobilities in the upstream cells. However, these mobilities are known, as all blocks upstream of the current cell have already been processed. For the purposes of the calculation therefore, FIN is a known constant.

FOUT is the rate of flow out of the cell to wells and to neighbouring cells with a lower potential. As before, potential differences are assumed to be fixed. The mobilities depend only on the saturation in the current (upstream) cell.

Equation (32) is therefore a non-linear equation in one unknown (the saturation in the current cell) which may be solved by any suitable algorithm.

Step 1 may be regarded as a first approximation, in which each phase is treated independently of the others. Step 2 is essentially a Newton correction, in which all of the phases are treated simultaneously to obtain a correction to the pressure field.

Because the cascade method is in effect a direct solution of a simplified problem, it has a number of important advantages:

(i) it can handle the advance of a saturation front across several cells in a single iteration;
(ii) non-linearities in the relative permeability curve can be handled on a cell by cell basis, and robust methods for solving non-linear equations in a single variable can be used;
(iii) it may be possible to use different grids and different time steps for the pressure and saturation equations. Thus the pressures might be solved using long time steps on a coarse grid, while saturations are solved using the cascade method on a much finer grid and with shorter time steps to minimise truncation errors.

However, the treatment of capillary pressure presents a problem for the cascade method, because it introduces a dependence of flow rate on the downstream saturation. Probably the simplest approach is to use capillary pressures computed at the previous iteration. Capillary pressure terms should however be fully expanded in the Jacobian used at each Newton iteration.

5. SOLUTION OF THE LINEAR EQUATIONS

The solution of the coupled non-linear equations using Newton's method involves the repeated solution of sets of linear equations. In the majority of simulations, it is this process which absorbs most of the computing time. This is particularly true for the long time steps often used in fully implicit simulations. The development of efficient solution methods is therefore of critical importance in the practical application of fully implicit methods to field-scale studies.

This section is devoted to a discussion of methods for the solution of the linear equations. The structure of the equations is described in Section 5.1, and this is followed by some observations concerning the effect on material balance of errors in the Jacobian. In Section 5.3, methods for decoupling the equations to allow sequential solution for pressures and saturations are reviewed. The remainder of the section is devoted to a discussion of preconditioned conjugate gradient methods, which currently offer the best performance on large problems. Multi-grid methods are a widely canvassed alternative to preconditioned conjugate gradient methods. However, these are as yet unproven in reservoir simulation applications and will not be discussed here. The interested reader is referred to the work of Behie and Forsyth,[12] and Brandt.[13]

For small problems, direct methods[14] are often competitive with (and sometimes superior to) iterative methods. However, storage and computational work increase very rapidly with problem size, making these methods impractical for field-scale studies.

5.1. The Structure of the Linear Equations

At each Newton iteration, we must solve a set of simultaneous linear equations of the form

$$Ax = b = -R \tag{33}$$

where

R is the residual vector	$\dfrac{\Delta M}{\Delta t} - F - Q$
A is the Jacobian matrix	$\dfrac{M'}{\Delta t} - F' - Q'$

and x (the solution) is an update to the state vector X.

The structure of A is shown in Fig. 5. The mass accumulation derivative, $M'/\Delta t$, contributes only to the diagonal band d. The flow derivative F' contributes both to the diagonal d and to the symmetrically placed pairs of side bands (l_1, u_1), (l_2, u_2) and (l_3, u_3). Note that because flow between two cells must be accounted for in the material balance equations of both

$$F_{ij} = -F_{ji} \tag{34}$$

we have

$$\frac{\partial F_{ij}}{\partial X_j} = -\frac{\partial F_{ji}}{\partial X_j} \tag{35}$$

which implies that for each element of F' on a side-band there is an equal

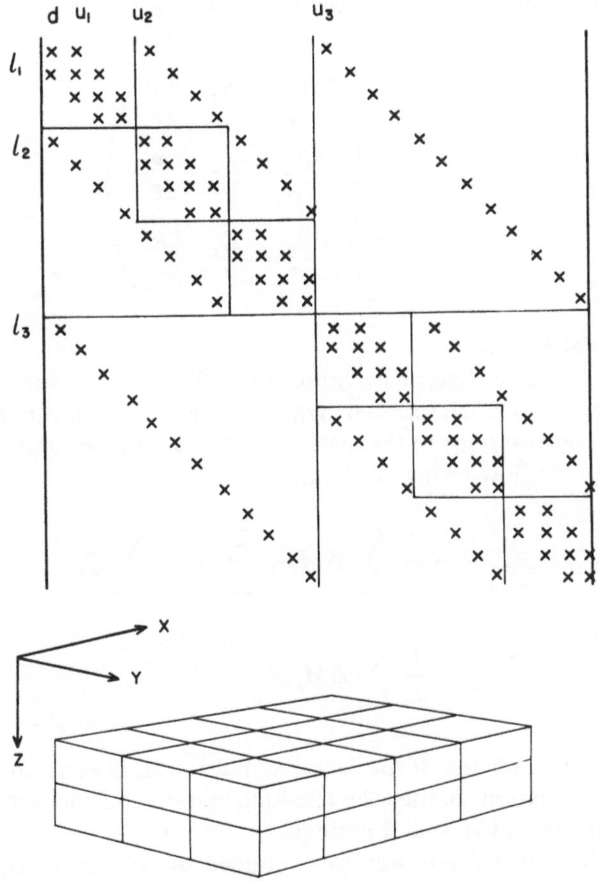

FIG. 5. The structure of the coefficient matrix from a simple three-dimensional problem. l_1, u_1 bands connect cells in the X-direction; l_2, u_2 in the Y-direction; l_3, u_3 in the Z-direction.

and opposite contribution to the diagonal in the same column. Thus the column sum of F' is zero. However, this does not imply symmetry, as in general

$$\frac{\partial F_{ij}}{\partial X_j} \neq \frac{\partial F_{ji}}{\partial X_i}$$

In most cases, the elements of A are not single numbers, but small matrices.

Thus, in a three-phase black oil system each element is a 3×3 matrix of the form

$$A = R' = \begin{pmatrix} \dfrac{\partial R_o}{\partial P} & \dfrac{\partial R_o}{\partial S_w} & \dfrac{\partial R_o}{\partial S_g} \\ \dfrac{\partial R_w}{\partial P} & \dfrac{\partial R_w}{\partial S_w} & \dfrac{\partial R_w}{\partial S_g} \\ \dfrac{\partial R_g}{\partial P} & \dfrac{\partial R_g}{\partial S_w} & \dfrac{\partial R_g}{\partial S_g} \end{pmatrix} \tag{36}$$

5.2. Material Balance

At any intermediate Newton iteration, eqn (29) will not be solved precisely and there will be a small residual error in each cell. The arithmetic sum of these errors corresponds to the material balance error per unit time, since, by eqn (34), the flow terms must cancel,

$$\begin{aligned} \sum_i R_i &= \frac{1}{\Delta t} \sum_i \Delta M_i - \sum_{i,n} F_{in} - \sum_i Q_i \\ &= \frac{1}{\Delta t} \sum_i \Delta M_i - \sum_i Q_i \end{aligned} \tag{37}$$

We will now show that if the Jacobian matrix, A, in eqn (33) is replaced by an approximation, B, then the resulting material balance error is zero if the column sums of A and B are equal.

If x' is the approximate solution obtained by solving the equation

$$Bx' = b \tag{38}$$

then the residual, or error, in the linear eqn (33) is

$$r = b - Ax' = (B - A)x' \tag{39}$$

Thus, the material balance error arising from use of the approximation will be zero if

$$\sum_i r_i = \sum_{i,j} (B - A)_{ij} x'_j = 0 \tag{40}$$

where the sums extend over all cells in the reservoir. Clearly, eqn (40) is satisfied for any value of x' if

$$\sum_i B_{ij} = \sum_i A_{ij} \tag{41}$$

that is if

$$\text{colsum}(B) = \text{colsum}(A) \tag{42}$$

where colsum(X) is the diagonal matrix formed by summing the elements of the matrix X in columns. This result, pointed out by Cheshire *et al.*,[3] is of considerable importance in the derivation of effective sequential and iterative methods for the solution of the linear eqn (33).

5.3. Decoupling the Phases

The simultaneous solution of the linear equations for all phases is a very time-consuming process, and a significant improvement in the efficiency of fully implicit simulators might be achieved if it were possible to decouple the phases, allowing sequential solution of the pressure and saturation

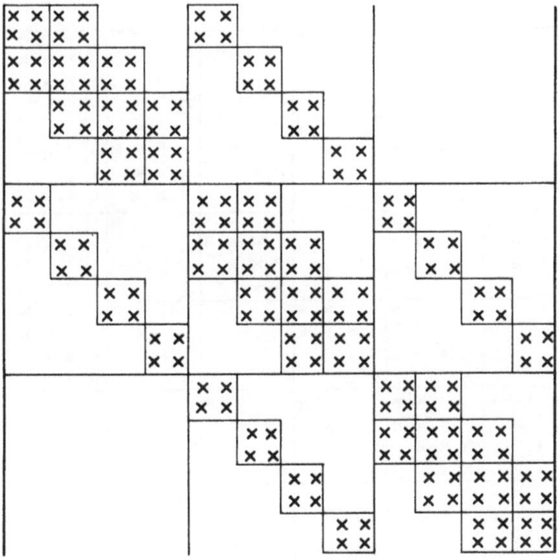

FIG. 6. Coefficient matrix for a two-phase two-dimensional (4 × 3) simulation. Each submatrix is a 2 × 2 matrix.

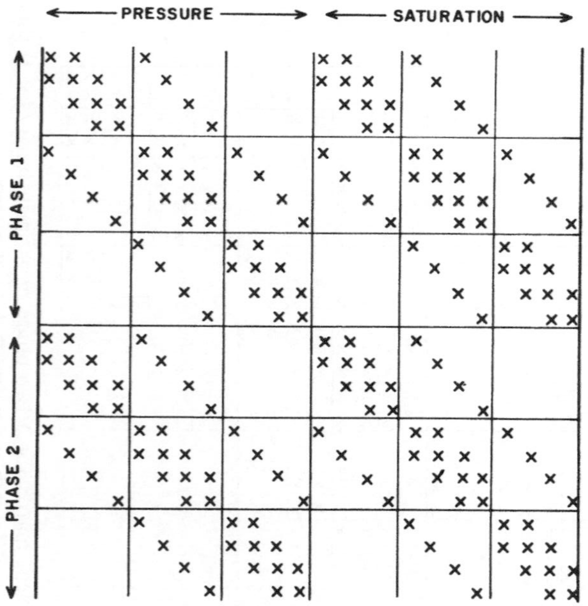

FIG. 7. Coefficient matrix for a two-phase two-dimensional (4 × 3) simulation.

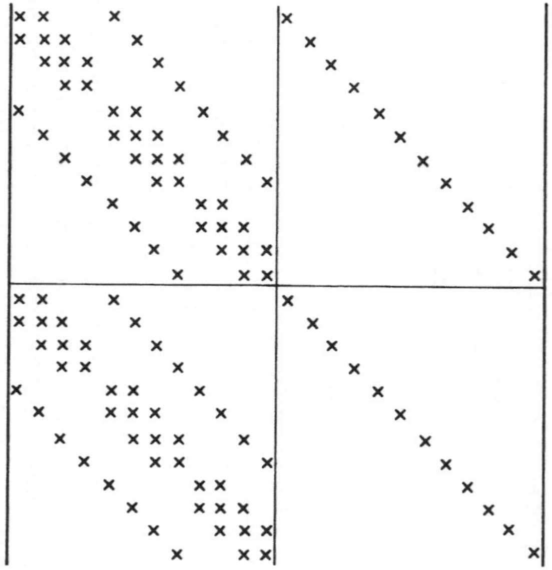

FIG. 8. Coefficient matrix for the IMPES method (cf. Fig. 7).

equations. In this section, we review some possible approaches to this goal. For simplicity, the methods are applied to a two-phase oil–water system. The principles are the same for more complex systems, although the algebra is more complicated.

As discussed in Section 5.1, the Jacobian matrix can be written as a banded matrix, each element of which is a 2×2 submatrix (see Fig. 6). However, it may also be written as a 2×2 matrix, each element of which is a banded matrix (see Fig. 7). Each of these banded matrices may be expressed as the sum of a diagonal matrix of mass accumulation derivatives M', and a banded matrix of flow derivatives F'.

The simplest decoupling algorithm is based on the observation that flow rates are strongly dependent on potential differences but only relatively weakly dependent on saturations. This suggests that a reasonable approximation to A is obtained by setting derivatives of flow with respect to saturation to zero (see Fig. 8).

$$B = \begin{pmatrix} \dfrac{\partial R_o}{\partial P} & \dfrac{1}{\Delta t}\dfrac{\partial M_o}{\partial S_w} \\ \dfrac{\partial R_w}{\partial P} & \dfrac{1}{\Delta t}\dfrac{\partial M_w}{\partial S_w} \end{pmatrix} \tag{43}$$

which can be factored for sequential solution by

$$B = \begin{pmatrix} I & G \\ 0 & I \end{pmatrix} \begin{pmatrix} \dfrac{\partial R_o}{\partial P} - G\dfrac{\partial R_w}{\partial P} & 0 \\ \dfrac{\partial R_w}{\partial P} & \dfrac{1}{\Delta t}\dfrac{\partial R_w}{\partial S_w} \end{pmatrix} \tag{44}$$

where G is the diagonal matrix $(\partial M_o/\partial S_w)(\partial M_w/\partial S_w)^{-1}$. It will be noted that, because G is diagonal, the matrix $(\partial R_o/\partial P) - G(\partial R_w/\partial P)$ has the same banded structure as $\partial R_o/\partial P$ and $\partial R_w/\partial P$. The above factorisation is the same as that used in the IMPES solution procedure.

A more accurate approximation to A is obtained by retaining the flow derivative in the second factor

$$B = \begin{pmatrix} I & G \\ 0 & I \end{pmatrix} \begin{pmatrix} \dfrac{\partial R_o}{\partial P} - G\dfrac{\partial R_w}{\partial P} & 0 \\ \dfrac{\partial R_w}{\partial P} & \dfrac{1}{\Delta t}\dfrac{\partial M_w}{\partial S_w} - \dfrac{\partial F_w}{\partial S_w} \end{pmatrix} \tag{45}$$

where

$$G = \left(\frac{\partial M_o}{\partial S_w}\right)\left(\frac{\partial M_w}{\partial S_w}\right)^{-1} \tag{46}$$

so that

$$B = \begin{pmatrix} \dfrac{\partial R_o}{\partial P} & \dfrac{1}{\Delta t}\dfrac{\partial M_o}{\partial S_w} - G\dfrac{\partial F_w}{\partial S_w} \\ \dfrac{\partial R_w}{\partial P} & \dfrac{1}{\Delta t}\dfrac{\partial M_w}{\partial S_w} - \dfrac{\partial F_w}{\partial S_w} \end{pmatrix} \tag{47}$$

This is the factorisation used in the standard sequential solution procedure. Unlike the IMPES factorisation, this approximation does not conserve material, because the column sum of B is not the same as that of the Jacobian, A. As a result, use of this factorisation gives rise to a persistent material balance error.

Recently, Cheshire *et al.*[3] have described a way in which the sequential procedure can be modified to ensure that material is conserved. Essentially, they show how the diagonal matrix, G, can be redefined so as to force the column sum of B to equal that of A:

$$B = \begin{pmatrix} I & G \\ 0 & I \end{pmatrix}\begin{pmatrix} \dfrac{\partial R_o}{\partial P} - G\dfrac{\partial R_w}{\partial P} & 0 \\ \dfrac{\partial R_w}{\partial P} & \dfrac{\partial R_w}{\partial S_w} \end{pmatrix}$$

$$= \begin{pmatrix} \dfrac{\partial R_o}{\partial P} & G\dfrac{\partial R_w}{\partial S_w} \\ \dfrac{\partial R_w}{\partial P} & \dfrac{\partial R_w}{\partial S_w} \end{pmatrix} \tag{48}$$

where G is determined by the condition that colsum(B) = colsum(A), so that

$$\text{colsum}\left(\frac{\partial R_o}{\partial S_w}\right) = \text{colsum}\left(G\frac{\partial R_w}{\partial S_w}\right) \tag{49}$$

It may be shown that a vector, $\mathbf{G}_v$, containing the diagonal elements of G is obtained by solving the matrix equation

$$\left(\frac{\partial R_w}{\partial S_w}\right)^{\mathrm{T}} \mathbf{G}_v = \text{colsum}\left(\frac{\partial R_o}{\partial S_w}\right) \tag{50}$$

where the right-hand side is a column vector containing the diagonal elements of colsum $(\partial R_o/\partial S_w)$.

In practice none of the above methods appears to be sufficiently robust to deal with large time steps in a fully implicit simulation. The problem appears to be that, whilst it is permissible to ignore almost all the terms in $\partial F_o/\partial S_w$ and $\partial F_w/\partial S_w$, there are a few specific terms, particularly in regions where the saturations are changing rapidly, which must be present if a robust method is to be obtained. The adaptive implicit method described by Thomas and Thurnau[5,6] tackles this requirement head on. Cells which require implicit treatment are identified using some simple criterion (e.g. those where the saturation changes are large) and the corresponding matrix elements are computed and stored. An example of the resulting matrix structure is shown in Fig. 9. It is now possible to decouple some of the equations (in cells where saturation changes are small) by a procedure similar to the IMPES factorisation. The remaining equations, which must be solved simultaneously, involve all the pressures and a few of the saturations. A full description of the procedure is given in Refs (5) and (6).

The adaptive implicit method probably gives a better approximation to

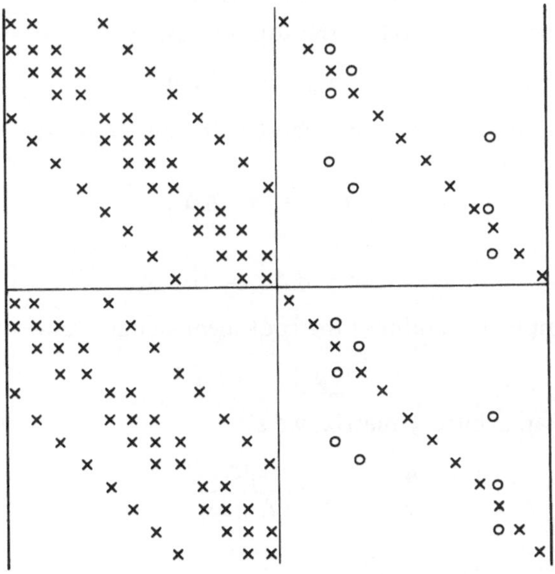

FIG. 9. Coefficient matrix for the adaptive implicit method. Flow derivatives are restored for selected saturations.

the complete Jacobian than any of the other decoupling methods. However it does lead to a very complicated matrix structure, which makes it difficult to program. In particular, the implementation of an iterative solver which fully exploits the reduction in the number of equations appears to be extremely difficult. The method will only be applicable to large-scale simulations if this difficulty can be overcome.

5.4. Conjugate Gradients

Although the method of conjugate gradients was first introduced by Hestenes and Stiefel[15] in 1952, it has only recently gained popularity in oil reservoir simulation. The method is important because it can be used to optimise the performance of any iterative procedure for solving the linear eqn (33). We will outline the main points of the technique below.

Any approximate solution x_i to eqn (33) gives rise to an error, or residual, defined by

$$r_i = b - Ax_i \tag{51}$$

All iterative methods depend on some approximation, B, to A, which is easily inverted to give

$$\Delta x_i = B^{-1} r_i \tag{52}$$

which can be used to update the solution using

$$x_{i+1} = x_i + \alpha_i \Delta x_i \tag{53}$$

The constant α_i may be chosen to minimise the residual r_{i+1} corresponding to x_{i+1}:

$$r_{i+1} = r_i - \alpha_i \Delta r_i \tag{54}$$

where

$$\Delta r_i = A\, \Delta x_i = AB^{-1} r_i \tag{55}$$

For example, to minimise the root-mean-square value of

$$s_{i+1} = Cr_{i+1} \tag{56}$$

where C is an arbitrary matrix, we set

$$\alpha_i = \frac{\langle s_i | \Delta s_i \rangle}{\langle \Delta s_i | \Delta s_i \rangle} \tag{57}$$

where

$$\begin{aligned} s_i &= Cr_i \\ \Delta s_i &= C\, \Delta r_i = Ms_i \end{aligned} \tag{58}$$

and

$$M = CAB^{-1}C^{-1} \tag{59}$$

This ensures that s_{i+1} is orthogonal to the search direction Δs_i, a condition which minimises $|s_{i+1}|^2$

$$\langle s_{i+1}|\Delta s_i\rangle = 0 \tag{60}$$

By including components of previous search directions into Δx_i and Δs_i

$$\Delta x_i = B^{-1}r_i + \sum_{l=1}^{i-1} \beta_{li}\,\Delta x_l$$

$$\Delta s_i = Ms_i + \sum_{l=1}^{i-1} \beta_{li}\,\Delta s_l \tag{61}$$

it is possible to ensure that $|s_{i+1}|^2$ is a global minimum over the i-dimensional subspace defined by the initial residual s_i and the set of search directions Δs_l. The choice of β_{li} ensures that the search directions are mutually orthogonal.

In principle, the above method guarantees convergence to a solution in n iterations (where n is the number of equations). In practice, this guarantee is of little value, as a practical method must converge in far fewer than n iterations, and rounding errors prevent even this target from being achieved. The practical value of the method depends on how well the preconditioning matrix, B, approximates to A.

It may be shown by induction, if M is symmetric, then $\beta_{li} = 0$ for $l < i-1$, so that each new search direction needs be orthogonalised only with respect to its predecessor. The algorithm described by Meijerink and Van der Vorst[7] relies on this property to accelerate convergence in problems for which the coefficient matrix is symmetric and positive definite. Thus if

$$A = LL^{\mathrm{T}} \tag{62}$$

where L is a matrix and L^{T} is its transpose and

$$C = L^{-1} \tag{63}$$

then

$$M = CAB^{-1}C^{-1} = L^{\mathrm{T}}B^{-1}L \tag{64}$$

which is symmetric if B is symmetric.

Unfortunately, multi-phase reservoir simulation always gives rise to asymmetric matrices. If the phases are solved simultaneously, the asymmetry is profound, and algorithms which depend on symmetry are very unlikely to work. If the phases are uncoupled, the matrix for the pressure equation is likely to be near-symmetric, and it may be possible to devise methods which exploit this fact.

One obvious way of ensuring that the coefficient matrix is symmetric and positive definite is to premultiply the equations by the transpose of coefficient matrix. Thus, instead of solving eqn (33), we solve

$$A^{T}Ax = A^{T}b \tag{65}$$

However, the resulting algorithm is usually rather slow, partly because the work per iteration is doubled, and partly because $B^{T}B$ is a weaker approximation to $A^{T}A$ than B is to A.

Usually, a better alternative is the orthomin procedure described by Vinsome.[8] This method does not rely on the zero value of β_{li} for $l < i-1$. Instead, orthogonalisation coefficients are computed explicitly for the previous κ search directions where κ is a parameter of the method. Thus at each iteration, the residual is minimised over a κ-dimensional subspace. The effectiveness of this procedure depends on the accuracy of the preconditioning matrix. With good preconditioning the solution is found in relatively few iterations and a small value of κ is sufficient to ensure near-optimal convergence. If, on the other hand, the preconditioning matrix is inaccurate, the iteration may not converge at all, particularly if κ is small.

In general, orthomin is applied with $C = I$, so that the procedure minimises the rms residual. If the preconditioning matrix is a structurally symmetric LU factorisation ($B = LU$), an interesting alternative is the choice $C = L^{-1}$ so that

$$M = L^{-1}AU^{-1} \tag{66}$$

The virtue of this choice is that if A and B are nearly symmetric then M will be too. As a result the orthogonalisation coefficients β_{li} should be small for $l < i-1$. However this possible gain must be set against a slightly increased initialisation cost, and it is not clear at present which factor is the more important.

5.5. The Preconditioning Matrix

The rate of convergence of preconditioned conjugate gradient methods depends critically on how well the preconditioning matrix, B, approximates the matrix A which, for three-dimensional problems can be written as

$$A = d + l_1 + u_1 + l_2 + u_2 + l_3 + u_3 \tag{67}$$

Most published preconditioning matrices are incomplete Cholesky decompositions of A. A complete Cholesky factorisation (as used in direct methods) generates upper and lower triangular factors which are much less sparse than the original matrix A (see Fig. 10). In an incomplete Cholesky factorisation (see Fig. 11), any terms which fall outside a predetermined sparsity pattern are ignored. As a result the incomplete factorisation is approximate, but can be computed much more quickly and with much less storage than a complete factorisation.

For two-dimensional problems (with $l_3 = u_3 = 0$) the simplest factorisation is one which preserves the sparsity of the original matrix

$$B = (\gamma + l_1 + l_2)\gamma^{-1}(\gamma + u_1 + u_2) \tag{68}$$

where γ is the diagonal matrix defined by

$$\gamma = d - l_1\gamma^{-1}u_1 - l_2\gamma^{-1}u_2 \tag{69}$$

Thus, $B = A + E$ where

$$E = l_1\gamma^{-1}u_2 + l_2\gamma^{-1}u_1 \tag{70}$$

This is the preconditioning matrix described as ICCG(0) in Ref. 7, but generalised to the case of asymmetric matrices.

Next in complexity comes an approximation with one additional band in each factor

$$B = (\gamma + m_1 + l_2 - l_2\gamma^{-1}v_1)\gamma^{-1}(\gamma + v_1 + u_2 - m_1\gamma^{-1}u_2)$$

where

$$\begin{aligned} m_1 &= l_1 + l_2\gamma^{-1}m_1\gamma^{-1}u_2 \\ v_1 &= u_1 + l_2\gamma^{-1}v_1\gamma^{-1}u_2 \\ \gamma &= d - m_1\gamma^{-1}v_1 - l_2\gamma^{-1}u_2 - l_2\gamma^{-1}v_1\gamma^{-1}m_1\gamma^{-1}u_2 \end{aligned}$$

so that

$$E = -(m_1\gamma^{-1}m_1\gamma^{-1}u_2 + l_2\gamma^{-1}v_1\gamma^{-1}v_1) \tag{71}$$

Additional bands may be included in the factors until the required degree of accuracy is achieved.

As increasingly accurate factorisations are constructed, it becomes economical to reorder the equations according to the D4 scheme of Price and Coats.[14] In fact these authors have shown that for complete Cholesky factorisation, reordering yields a gain of a factor of two or more. This principle has been applied to incomplete factorisation by Tan and Letkeman.[16] It should however be noted that D4 ordering yields a less accurate approximation if no additional bands are included in the factors,

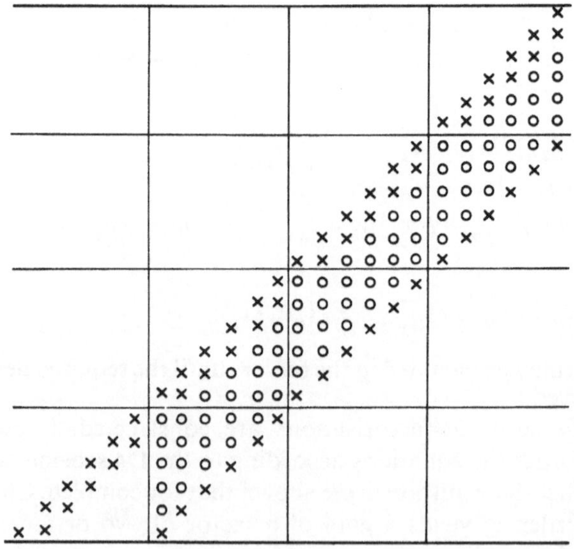

FIG. 10. Lower triangular factor of a complete Cholesky decomposition of a two-dimensional (6 × 4) problem: × denotes elements with same sparsity as the original matrix. ○ denotes additional elements required for Gaussian elimination.

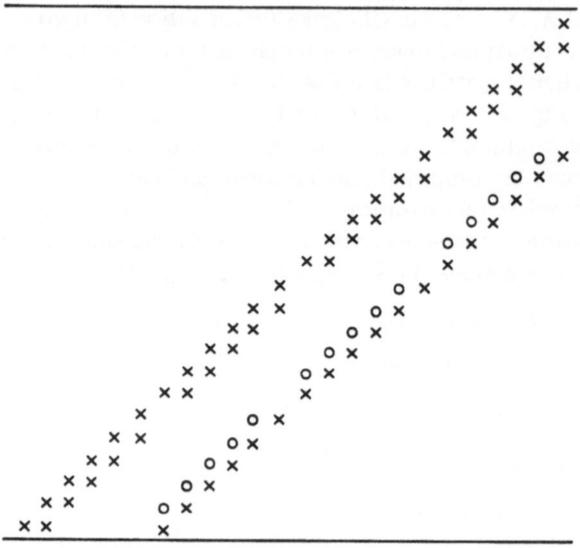

FIG. 11. Sparsity pattern for an incomplete Cholesky factorisation.

and that the advantage conferred by the scheme is apparent only when the factorisation is more complete. Watts[17] has described a scheme in which the equations are ordered according to the D2 scheme, which also confers an advantage over natural ordering if there are enough additional bands.

Incomplete Cholesky factorisations can be modified to ensure that the column sum of the error matrix is zero. The simplest approximation of this type is

$$\begin{aligned} B &= (\gamma + l_1 + l_2)\gamma^{-1}(\gamma + u_1 + u_2) \\ \gamma &= d - l_1\gamma^{-1}u_1 - l_2\gamma^{-1}u_2 - \text{colsum}\,(l_1\gamma^{-1}u_2 + l_2\gamma^{-1}u_1) \\ E &= l_1\gamma^{-1}u_2 + l_2\gamma^{-1}u_1 - \text{colsum}\,(l_1\gamma^{-1}u_2 + l_2\gamma^{-1}u_1) \end{aligned} \tag{72}$$

The object of this modification is to force the sum of the residuals, which in a reservoir simulator corresponds to a material balance error, to zero. If the iteration is started from the initial solution

$$x_o = B^{-1}b \tag{73}$$

then the initial residual r_o is given by

$$r_o = b - Ax_o = (B - A)B^{-1}b \tag{74}$$

and the sum of the elements of r_o is zero if colsum $(B - A) = 0$. This condition, once established, remains valid throughout the iteration. Gustafsson[18] suggests that methods modified as described above converge in $0(n^{5/4})$ operations as opposed to $0(n^{3/2})$ for unmodified methods. He has managed to prove this result for some simple cases and presented numerical evidence for others. Appleyard *et al.*[4] and Behie and Forsyth[12] have also provided evidence that modified methods are superior, particularly when the number of additional bands is small.

5.6. Nested Factorisation

A new preconditioning method, which is not an incomplete Cholesky factorisation, has recently been proposed by Appleyard *et al.*[4] The method appears to compare very favourably with ICCG[7] methods in both computing time and storage requirements.

The essential features of the method as applied to two-dimensional problems are shown in Fig. 12. The approximation is formed as the product of factors, but, unlike an incomplete Cholesky factorisation (see Fig. 11) the factors are not strictly lower and upper triangular. Instead a block triangular structure is obtained by including one band from the upper matrix in the lower factor, and one from the lower matrix in the upper

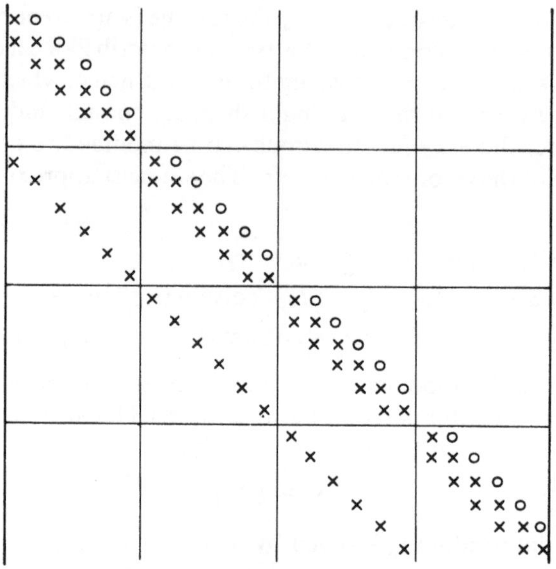

FIG. 12. The lower factor of a nested factorisation. Note the presence of an upper band (denoted by ○).

factor. Each factor has the same shape as the approximation used in a Line Gauss Seidel iteration.

$$B = (\beta + l_2)\beta^{-1}(\beta + u_2) \tag{75}$$

The matrix β is tridiagonal and, as such, can be expressed as the product of strictly lower, diagonal and upper factors:

$$\beta = (\gamma + l_1)\gamma^{-1}(\gamma + u_1) \tag{76}$$

The diagonal matrix γ is determined by the requirement that the column sum of the error matrix be zero:

$$\gamma = d - l_1\gamma^{-1}u_1 - \text{colsum}\,(l_2\beta^{-1}u_2) \tag{77}$$

Combining these equations, we obtain an expression for B, i.e.

$$B = A + l_2\beta^{-1}u_2 - \text{colsum}\,(l_1\beta^{-1}u_2) \tag{78}$$

A comparison between this approximation and incomplete Cholesky factorisations is obtained by combining eqns (75) and (76) to express B as the product of strictly triangular factors:

$$B = [\gamma + l_1 + l_2(I + \gamma^{-1}u_1)^{-1}]\gamma^{-1}[\gamma + u_1 + (I + l_1\gamma^{-1})^{-1}u_2] \tag{79}$$

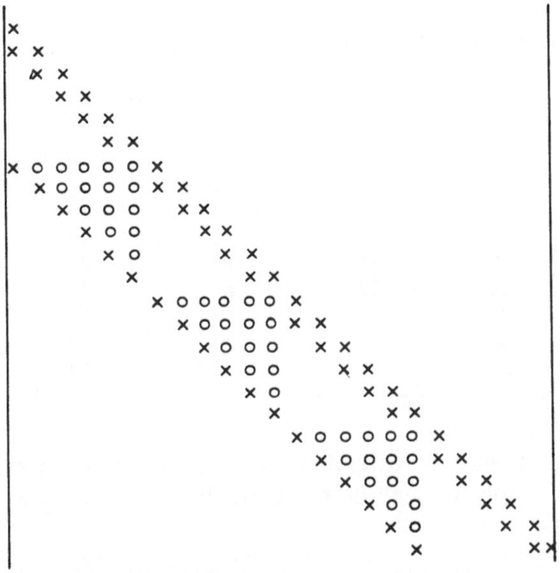

FIG. 13. Equivalent Cholesky representation of the lower factor in a nested factorisation.

The sparsity pattern of the lower factor is shown in Fig. 13. It can be seen that to achieve the same accuracy using an incomplete Cholesky factorisation it would be necessary to include a large number of additional bands in the factors—a process which adds to both the computing time and storage requirements of the algorithm.

Table 1 shows how the nested factorisation algorithm compares with a number of alternatives on a set of two-dimensional problems described by Settari and Aziz.[19] For each problem, the table shows the computational work required to reduce the largest normalised residual to 10^{-6}. Further details are given in Ref. 20.

The results show that nested factorisation is fastest of the methods tested on all the problems. Perhaps the most significant result is that on problem 2, which is the only one to exhibit the strong directionality characteristic of cross-sectional reservoir simulations. On this problem nested factorisation is fastest by a factor of three. However, in this case it is important to align the axes so that the l_1 and u_1 bands contain the largest matrix elements, as convergence takes about twice as long if the axes are wrongly aligned. In general, it is worthwhile to include code which selects the correct alignment automatically.

TABLE 1
COMPARISON OF NESTED FACTORISATION WITH OTHER ITERATIVE METHODS

Problem number	*Computational work (SIP iterations)*				
	SIP	*ICCG0*	*ICCG0 (with colsum)*	*ICCG3*	*Nested factorisation*
1	28	30	18	19	12
2	20	24	19	27	6
3	38	33	21	21	12
4	28	29	19	19	13
5	>50	32	20	20	11
6	50	25	16	14	9

The simplest way of extending the algorithm to three-dimensional problems is simply to include the extra bands in the lower and upper factors:

$$\begin{aligned} B &= (\beta + l_2 + l_3)\beta^{-1}(\beta + u_2 + u_3) \\ \beta &= (\gamma + l_1)\gamma^{-1}(\gamma + u_1) \\ \gamma &= d - l_1\gamma^{-1}u_1 - \text{colsum}\,[(l_2 + l_3)\beta^{-1}(u_2 + u_3)] \end{aligned} \tag{80}$$

In practice, this approximation often proves inadequate for difficult three-dimensional problems, and a different procedure which exploits the block tridiagonal structure of the matrix is to be preferred:

$$\begin{aligned} B &= (\alpha + l_3)\alpha^{-1}(\alpha + u_3) \\ \alpha &= (\beta + l_2)\beta^{-1}(\beta + u_2) \\ \beta &= (\gamma + l_1)\gamma^{-1}(\gamma + u_1) \\ \gamma &= d - l_1\gamma^{-1}u_1 - \text{colsum}\,(l_2\beta^{-1}u_2 + l_3\alpha^{-1}u_3) \end{aligned} \tag{81}$$

so that

$$B = A + l_2\beta^{-1}u_2 + l_3\alpha^{-1}u_3 - \text{colsum}\,(l_2\beta^{-1}u_2 + l_3\alpha^{-1}u_3) \tag{82}$$

This again approximates A by the product of block triangular factors (rather than the strictly triangular incomplete Cholesky factors). The block diagonal matrix α, which connects cells within a plane, is itself the product of block triangular factors. In fact, α has the same structure as the approximation used for two-dimensional problems.

An important advantage of the nested factorisation algorithm is its very modest storage requirements. Both two- and three-dimensional versions require storage for only two diagonal matrices (γ^{-1} and $\gamma^{-1}u_1$) in addition to the elements of A and vectors required for the conjugate gradient acceleration. Incomplete Cholesky factorisations generally require much more storage, particularly for three-dimensional problems. As an example, Watts' algorithm[17] requires storage for 15 diagonal matrices in two-dimensions or 35 for three-dimensional problems.

The evaluation of γ^{-1} proceeds a plane at a time, and within each plane, a line at a time. At each stage the calculation involves elements of γ^{-1} from the previous cell in the current line, the previous line in the current plane and the previous plane. The evaluation of colsum $(l_2\beta^{-1}u_2 + l_3\alpha^{-1}u_3)$ is achieved by multiplying a column vector whose elements are all unity by the transpose of $(l_2\beta^{-1}u_2 + l_3\alpha^{-1}u_3)$. The resulting vector contains the diagonal elements of colsum $(l_2\beta^{-1}u_2 + l_3\alpha^{-1}u_3)$.

It will be noted that, as with the two-dimensional algorithm, γ has been set in such a way that the column sum of the error matrix $B - A$, is zero. As noted previously, this ensures that the sum of the residuals (material balance error) is zero throughout the iteration. In this case, however, the error matrix is block diagonal, with the result that the residuals sum to zero independently within each plane (or, for two-dimensional problems, within each line). This fact may be used to verify the correctness of the implementation.

Whilst it is convenient when describing the algorithm to view it as a series of nested LDU factorisations (eqns (81)) a significantly more efficient implementation is obtained by combining the D and U factors

$$
\begin{aligned}
B &= (\alpha + l_3)(I + \alpha^{-1}u_3) \\
\alpha &= (\beta + l_2)(I + \beta^{-1}u_2) \\
\beta &= (\gamma + l_1)(I + \gamma^{-1}u_1)
\end{aligned}
\tag{83}
$$

In addition, the most rapid convergence is obtained if the axes are aligned so that the largest matrix elements fall on the l_1 and u_1 bands, the next largest on the l_2 and u_2 bands, and the smallest on the l_3 and u_3 bands. In a reservoir simulator, a reasonable procedure appears to use the sum of the transmissibilities along the three directions to align the axes.

In Table 2, the three-dimensional nested factorisation algorithm is compared with the ICCG algorithm of Watts for the series of three-dimensional test problems described in Ref. 17. As before, the table shows the computational work required to reduce the maximum normalised

TABLE 2
COMPARISON OF NESTED FACTORISATION WITH THE METHOD OF WATTS[17]

Problem number	*Dimensions*	*Computational work (SIP iterations)*	
		Watts	*Nested factorisation*
4	21 × 21 × 21	45	20
5	21 × 21 × 21	60	11
6	21 × 21 × 21	102	16
7	21 × 21 × 5	47	11
8	21 × 21 × 5	50	16

residual to 10^{-6}. Nested factorisation performs better on all these problems. However, it is noticeable that, unlike Watts' method, nested factorisation performs better on problems with one or two strong directions (Nos 5, 6 and 7) than on isotropic problems (No. 4), or problems with mixed directionality (No. 8).

REFERENCES

1. Bansal, P. P., Harper, J. L., McDonald, A. E., Moreland, E. E. and Odeh, A. S. (1979) A strongly coupled, fully implicit, three dimensional, three phase reservoir simulator. Paper SPE 8329, *SPE-AIME 54th Annual Fall Meeting of the SPE*, Las Vegas, Nevada.
2. Au, A. D. K., Behie, A., Rubin, B. and Vinsome, P. K. W. (1980) Techniques for fully implicit reservoir simulation. Paper SPE 9302, *SPE-AIME 55th Annual Fall Meeting of the SPE*, Dallas, Texas.
3. Cheshire, I. M., Appleyard, J. R., Banks, D., Crozier, R. J. and Holmes, J. A. (1980) An efficient fully implicit simulator. Paper EUR 179, *European Offshore Petroleum Conference and Exhibition*, London.
4. Appleyard, J. R., Cheshire, I. M. and Pollard, R. K. (1981) Special techniques for fully-implicit simulators. *Proceedings of the European Symposium on Enhanced Oil Recovery*, Elsevier Science Publishing Co., Amsterdam, pp. 395–408.
5. Thomas, G. W. and Thurnau, D. H. (1981) Reservoir simulation using an adaptive implicit method. Paper SPE 10120, *SPE-AIME 56th Annual Fall Meeting of the SPE*, San Antonio, Texas; *Soc. Pet. Eng. J.* (1983), **23**(5), 759–68.
6. Thomas, G. W. and Thurnau, D. H. (1982) The mathematical basis of the adaptive implicit method. Paper SPE 10495, *Sixth SPE Symposium on*

Reservoir Simulation, New Orleans, Louisiana; *Soc. Pet. Eng. J.* (1983), **23**(5), 759–68.

7. MEIJERINK, J. A. and VAN DER VORST, H. A. (1977) An iterative solution method for linear systems of which the coefficient matrix is a symmetric M-matrix. *Mathematics of Computation*, **31** (January), 148–62.
8. VINSOME, P. K. W. (1976) Orthomin, an iterative method for solving sparse banded sets of simultaneous linear equations. Paper SPE 5729, *SPE-AIME Fourth Symposium on Reservoir Simulation*, Los Angeles, California.
9. AZIZ, K. and SETTARI, A. (1980) *Petroleum Reservoir Simulation*, Elsevier Applied Science Publishers, London.
10. BANKS, D. and PONTING, D. K. (1981) Interphase mass transfer effects in implicit black oil simulators. *Proceedings of the European Symposium on Enhanced Oil Recovery*, Elsevier Science Publishing Co., Amsterdam.
11. CHAPPELEAR, J. E. and ROGERS, W. L. (1974) Some practical considerations in the construction of semi-implicit simulators. *Soc. Pet. Eng. J.*, **14**(3), 216–20.
12. BEHIE, A. and FORSYTH, P. A. (1982) Multi-grid solution of the pressure equation in reservoir simulation. Paper SPE 10492, *Sixth SPE Symposium on Reservoir Simulation*, New Orleans, Louisiana; *Soc. Pet. Eng. J.* (1983), **23**(4), 623–32.
13. BRANDT, A. (1977) Multi-level adaptive solutions to boundary-value problems. *Mathematics of Computation*, **31** (April), 333–90.
14. PRICE, H. S. and COATS, K. H. (1974) Direct methods in reservoir simulation. *Soc. Pet. Eng. J.*, **14**(3), 295–308.
15. HESTENES, M. R. and STIEFEL, E. (1952) Method of conjugate gradients for solving linear systems. *J. Res. Nat. Bur. Standards*, **42**, 409–36.
16. TAN, T. B. S. and LETKEMAN, J. P. (1982) Application of D4 ordering and minimisation in an effective partial matrix inverse method. Paper SPE 10493, *Sixth SPE Symposium on Reservoir Simulation*, New Orleans, Louisiana.
17. WATTS, J. W. (1981) A conjugate gradient–truncated direct method for the iterative solution of the reservoir simulation pressure equation. *Soc. Pet. Eng. J.*, **21**(3), 345–53.
18. GUSTAFSSON, I. (1978) A class of first order factorisation methods. *BIT* **18** (April) 142–56.
19. SETTARI, A. and AZIZ, K. (1973) A generalisation of the additive correction methods for the iterative solution of matrix equations. *Siam J. Numer. Anal.* **10** (June) 506–21.
20. APPLEYARD, J. R. and CHESHIRE, I. M. (1983) Nested factorisation. Paper SPE 12264, *Seventh SPE Symposium on Reservoir Simulation*, San Francisco, California.

Chapter 8

THERMAL PROPERTIES OF RESERVOIR ROCKS AND FLUIDS

M. J. PASSMORE*

Department of Mineral Resources Engineering, Royal School of Mines, Imperial College of Science and Technology, London, UK

and

JOHN S. ARCHER

Department of Mineral Resources Engineering, Royal School of Mines, Imperial College of Science and Technology, and ERC Energy Resource Consultants Ltd, London, UK

NOMENCLATURE

A', B', C', D'	Constants for a particular saturating fluid	—
A	Constant	—
B	Constant	—
B_o	Oil formation volume factor	—
C	Specific heat	$L^2t^{-2}T^{-1}$
c_f	Pore volume compressibility	Lt^2m^{-1}
D	Thermal diffusivity	L^2t^{-1}
D_ph_p/λ_m	Dimensionless consolidation parameter	—
f	Volume fraction of non-condensible gas in gas phase	—
F	Formation resistivity factor	—

* Present address: Exploration Consultants Ltd, Highlands Farm. Greys Road, Henley-on-Thames, Oxon RG9 4PS, UK.

k	Permeability	L^2
L_V	Latent heat of vaporisation	L^2t^{-2}
M_R	Volumetric heat capacity of reservoir	$mL^{-1}t^{-2}T^{-1}$
S	Saturation	—
t	Time	t
T	Temperature	T
ΔT	Temperature difference	T
V_p	Pore volume	L^3
Z_p	Compressional velocity	LT^{-1}
β	Volumetric thermal expansion coefficient	T^{-1}
γ	Specific gravity	—
ε	Conductivity factor	—
η	Ratio of effective thickness of fluid film adjacent to contact surface of two solid particles in the original packed bed	—
λ	Thermal conductivity	$mLt^{-3}T^{-1}$
μ	Viscosity	$mL^{-1}t^{-1}$
ρ	Density	mL^{-3}
ϕ	Porosity	—
ϕ'	Porosity of sandstone	—
ω	Isothermal expansion	—

Superscript

0	Stagnant phase

Subscripts

a	Air
b	Bulk
dr	Drainage
e	Effective
f	Fluid
g	Gas
imb	Imbibition
m	Matrix
n	Liquid-saturated rock
o	Oil
ref	Reference
r	Rock
s	Steam
T	Temperature
w	Water

1. INTRODUCTION

The two main methods for heavy oil recovery which have economic possibilities are steam injection and *in-situ* combustion.[1] Each method involves the introduction of heat into the reservoir and results in a number of displacement mechanisms, e.g.

viscosity reduction
distillation
solvent extraction
thermal expansion of oil
hot water drive
reduction in residual oil saturation

These mechanisms depend particularly on the physical properties of the reservoir rock and fluids and their variation with temperature and pressure.

The petroleum literature provides data for many of the properties of fluids. For example, properties of water and steam can be found in steam tables,[2] and many of the properties of hydrocarbons are given in texts such as Standing[3] and others.[1,4–6]

2. ROCK AND FLUID PROPERTIES

In this section the following rock and fluid properties are described: heat capacity, thermal conductivity, compressibility, thermal expansion, viscosity and petrophysical properties such as porosity, permeability, capillary pressure, and relative permeability.

2.1. Heat Capacity

Heat capacity can be defined as the amount of heat required to raise the temperature of a body through one degree Kelvin. When determined experimentally, heat capacity measurements are considered reproducible to within $\pm 0{\cdot}5$ per cent. Heat capacity can also be calculated using Kopp's Law which states that the heat capacity of a compound is equal to the sum of the heat capacities of its constituent elements. Somerton[7] tested this law for complex mixtures, such as sandstone and shales, by comparing calculated values with experimental results and found a maximum deviation of less than 2%. Based on this close agreement it has been postulated that the heat content of fluid-saturated porous rock may be calculated in a similar manner. Somerton demonstrated the effect of

TABLE 1

CALCULATED HEAT CAPACITIES OF FLUID-SATURATED ROCKS[7]

Sample	Heat capacity (in Btu/ft³/°F)[a] at 620°F and pressure indicated: 14·7 psia			500 psia		1 500 psia		3 000 psia	
	Dry	*Methane*	*Water*	*Methane*	*Water*	*Methane*	*Water*	*Methane*	*Water*
Sandstone	34·0	34·0	34·0	34·1	34·1	34·3	34·9	34·6	45·9
Silty sand	35·6	35·6	35·6	35·7	35·7	36·0	36·5	36·3	49·1
Siltstone	32·0	32·0	32·0	32·1	32·1	32·5	33·2	32·9	48·7
Shale	39·6	39·6	39·6	39·6	39·6	39·7	40·0	39·9	40·1
Limestone	35·4	35·4	35·4	35·5	35·5	35·7	36·2	36·0	46·8

[a] $1\ Btu/ft^3/°F \times 67{\cdot}07 = 1\ kJ/m^3/K$.

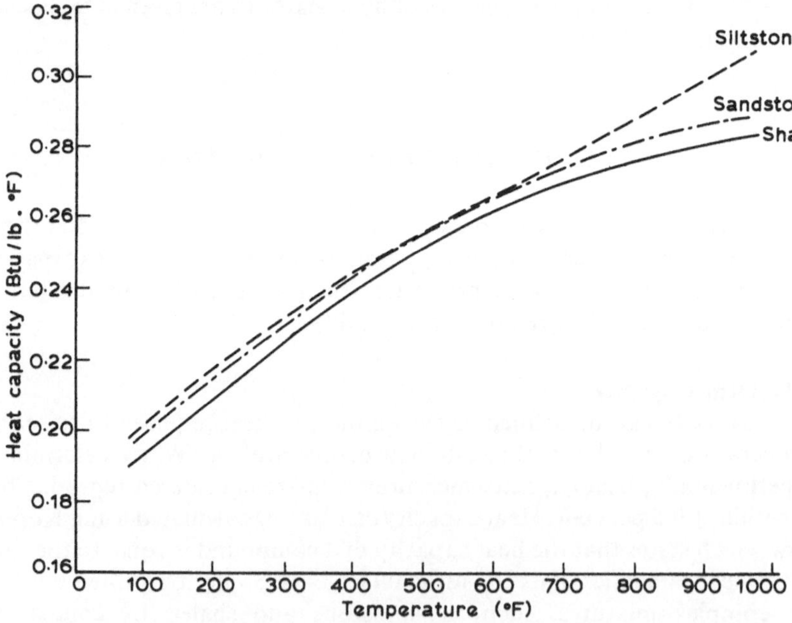

FIG. 1. Experimental heat capacities ($1\ Btu/lb\,.\,°F \times 4{\cdot}1868 = 1\ kJ/kg\,.\,K$).[7]

varying fluid saturation on the heat capacity of rock samples at various pressures and his results are shown in Table 1. It can be seen that gas saturation has negligible effect on the heat capacity of a porous rock but the presence of liquid, e.g. water, may increase the heat capacity by 35% or more depending on the pressure and temperature. The volumetric heat capacity of a fluid-saturated porous rock can usually be evaluated using the following expression:

$$M_R = (C\rho)_T = (1-\phi)C_r\rho_r + \phi(S_o\rho_oC_o + S_w\rho_wC_w) + \phi S_g\left[f\rho_gC_g + (1-f)\left(\frac{\rho_sL_V}{\Delta T}\right) + \rho_sC_w\right] \quad (2.1.1)$$

This equation has a contribution from up to four components—rock, oil, water and gas. Furthermore, the gas phase has two parts—one for condensible gas and one for non-condensible gas.

The variation of heat capacity measured for dry rock with temperature is illustrated in Fig. 1. The maximum deviation between any of the samples is less than 10% in this plot.

Heat capacity data and correlations for crude oils have been reported by Kesler and Lee[8] and others[4,6] and are based on API gravity and the Watson characterisation factor K_w. The effect of crude oil rather than water fluid in a pore space leads to only about half of the increase in heat capacity of the system.

2.2. Thermal Conductivity of Reservoir Rocks

Heat transfer in a porous rock saturated with fluid can occur by thermal conduction (a) in the grains, (b) in the fluid, (c) between grains and (d) between grain and fluid. The effective thermal conductivity of a porous system depends on the thermal conductivity of the solid matrix, the conductivity of the saturating phase and the phase saturation as a fraction of the pore volume.

Conceptual models and statistical correlations have been used to determine effective thermal conductivity, λ_e, in terms of known parameters. Conceptual models have tended to represent the phases either in series or parallel. The series distribution is given by

$$\frac{1}{\lambda_e} = \frac{\phi}{\lambda_f} + \frac{(1-\phi)}{\lambda_m} \quad (2.2.1)$$

The parallel distribution is given by

$$\lambda_e = \phi\lambda_f + (1-\phi)\lambda_m \quad (2.2.2)$$

The general forms of these equations give the weighted harmonic and arithmetic means respectively,

$$\frac{1}{\lambda_e} = \sum_{i=1}^{n} \frac{V_i}{\lambda_i} \tag{2.2.3}$$

$$\lambda_e = \sum_{i=1}^{n} V_i \lambda_i \tag{2.2.4}$$

which correspond to the minimum and maximum or bounding values of the probable effective thermal conductivity. The probability that the phases are in series or in parallel is equal in most systems. This can be represented by the weighted geometric mean which gives the intermediate values of the effective thermal conductivity, i.e.

$$\lambda_e = \prod_{i=1}^{n} \{\lambda_i\}^{V_i} \tag{2.2.5}$$

Kunii and Smith[9] developed an analytical equation to predict the conductivity of porous rock:

$$\frac{\lambda_e^0}{\lambda_m} = \phi'\left(\frac{\lambda_g}{\lambda_m}\right) + \frac{(1-\phi)\left(1+\dfrac{\phi}{\phi'}t\right)}{1+\dfrac{\phi/\phi'}{\dfrac{1}{\eta}\left(\dfrac{\lambda_g}{\lambda_m}+\dfrac{D_p h_p}{\lambda_m}\right)}} \tag{2.2.6}$$

where $D_p h_p/\lambda_m$ is described as a dimensionless consolidation parameter. In order to apply this equation in finding the thermal conductivity of porous rocks filled with stagnant fluid, λ_e^0, both the thermal conductivity of the solid, λ_m, and $D_p h_p/\lambda_m$ must be known. These are generally obtained by experiment. Kunii and Smith developed correlations for some sands and used eqn (2.2.6) to demonstrate that the effects of properties of the rock and fluid, e.g. void fraction and fluid conductivity, can be accounted for satisfactorily.

Zierfuss and van der Vliet[10] measured the thermal conductivity of 50 samples using a steady-state method. When arranged in order of increasing porosity the conductivity tended to decrease, although some exceptions were noticed. Typically the conductivity decreased from 2·51 Btu/h . ft . °F*

* 1 Btu/h . ft . °F × 6·230 = 1 kJ/h . m . K.

TABLE 2
COEFFICIENTS OF REGRESSION EQUATIONS[10]

Rock type	*Saturant*	*A′*	*B′*	*C′*	*D′*
Sandstone	Water	+3·874	−15·29	+31·84	−27·00
	Oil	+3·840	−15·75	+32·37	−27·79
	Gas	+3·860	−17·10	+33·60	−26·35
Limestone	Water	+2·470	−0·03	—	—
	Oil	+2·422	−0·03	—	—
	Gas	+2·356	−0·03	—	—

at a porosity of 17% to 0·925 Btu/h.ft.°F at 37·4%. Their original aim was the development of a method to find porosities from measured thermal conductivities in a bore hole. This was complicated by the influence of permeability on thermal conductivity for rocks of the same porosity. The observation of increased thermal conductivity with increased permeability was explained in terms of convective heat transport in wider pores. Since permeability is not easily measured *in situ* a compensation was achieved by defining the product of thermal conductivity (λ) and formation resistivity factor, F, which decreases as pore size increases. Multiple regression analysis of their results produced the following equation:

$$\log_{10} F\lambda = A' + B'\phi + C'\phi^2 + D'\phi^3 \qquad (2.2.7)$$

where A', B', C', D' are constants for a particular saturating fluid. Typical values for the constants are given in Table 2.

Assad[11] proposed the correlation

$$\frac{\lambda}{\lambda_m} = \frac{\lambda F^n}{\lambda_m} \qquad (2.2.8)$$

where $n = c\phi$; c is a correlation factor $\simeq 1$ and ϕ = fractional porosities.

Subsequently, Anand *et al.*[12] employed similar multiple regression analysis using commonly available physical properties of rocks in a relationship for predicting thermal conductivity:

$$\lambda_{dry} = 0{\cdot}340\rho - 3{\cdot}20\phi + 0{\cdot}53k^{0{\cdot}1} + 0{\cdot}013F - 0{\cdot}031 \qquad (2.2.9)$$

Of their 38 data points, agreement was within 15% for 31 of them. This is shown in Fig. 2. Considering the variables in eqn (2.2.9) porosity is the most important and density and permeability have an equal effect.

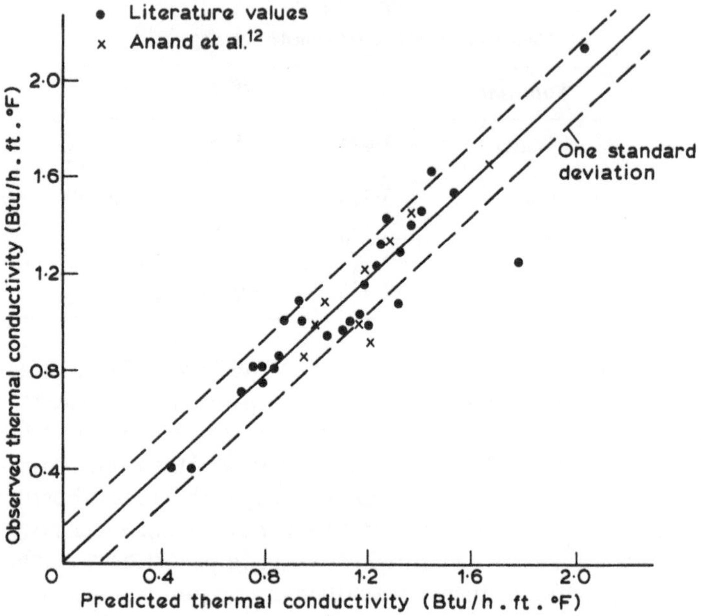

FIG. 2. Agreement between measured and calculated thermal conductivity values.[12]

For liquid-saturated sandstone, using dimensionless groupings the following equation from 52 data points was obtained:

$$\frac{\lambda_e}{\lambda_m} = 1{\cdot}00 + 0{\cdot}3\left[\frac{\lambda_f}{\lambda_a} - 1{\cdot}0\right]^{0\cdot 33} + 4{\cdot}57\left[\frac{\phi}{(1-\phi)}\frac{\lambda_f}{\lambda_m}\right]^{0\cdot 48m}\left[\frac{\rho_m}{\rho_r}\right]^{-4\cdot 3} \qquad (2.2.10)$$

As shown in Fig. 3, 85 % of the values showed agreement within 15 %. In a similar approach Evans[13] made a study of 39 rock samples from various parts of the North Sea, and obtained the following relationship:

$$\lambda = -0{\cdot}085\phi - 0{\cdot}91Z_p + 389\rho_b - 9{\cdot}52 \qquad (2.2.11)$$

The results shown in Fig. 4 indicated an agreement with experimentally measured thermal conductivity to within about 15 %.

2.2.1. *Temperature Dependence of Thermal Conductivity*

Thermal conductivities of most natural rock materials decrease with increased temperature up to about 1200 °F, when conductivity becomes

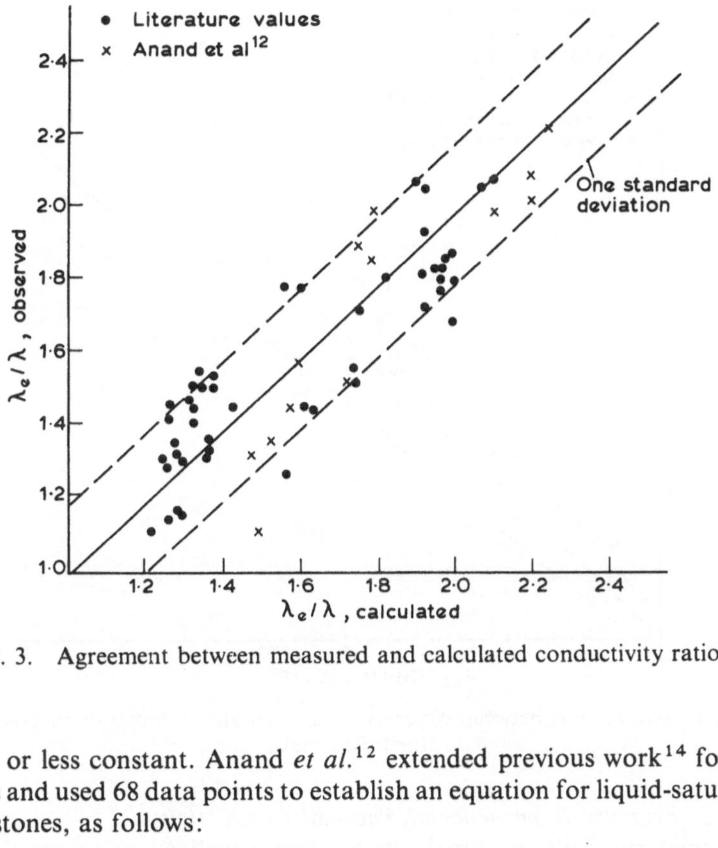

FIG. 3. Agreement between measured and calculated conductivity ratios.[12]

more or less constant. Anand *et al.*[12] extended previous work[14] for dry rocks and used 68 data points to establish an equation for liquid-saturated sandstones, as follows:

$$\lambda_T = \lambda_{68} - 0{\cdot}71 \times 10^{-3}(T - 528)(\lambda_{68} - 0{\cdot}8)[\lambda_{68}(T \times 10^{-3})^{-0{\cdot}55\lambda_{68}} + 0{\cdot}74] \tag{2.2.12}$$

where T is in degrees Rankine and λ_{68} is the thermal conductivity at 68 °F

Somerton *et al.*[15] developed a similar equation:

$$\lambda_t = \lambda_{125} - 1{\cdot}28 \times 10^{-3}(T - 125)(\lambda_{125} - 0{\cdot}82) \tag{2.2.13}$$

where λ_{125} is the thermal conductivity at 125 °F.

The variation of thermal conductivity with temperature predicted using these equations is shown in Fig. 5. These equations may need to be modified when applied to a wider range of rock-fluid systems, e.g. variations in porosity may have an effect, but they can be used as a general guide.

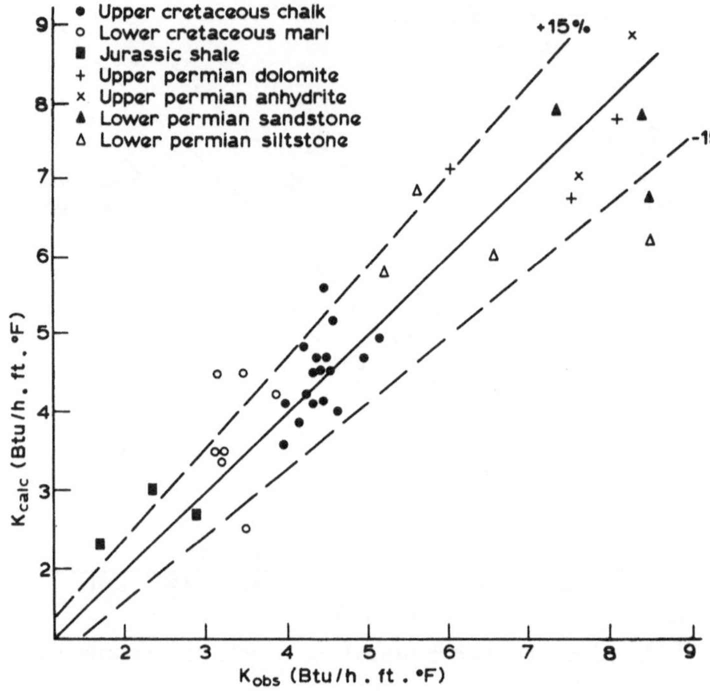

FIG. 4. Agreement between observed and calculated thermal conductivities for North Sea rocks.[13]

2.2.2. Pressure Dependence of Thermal Conductivity

Thermal conductivity increases with increase in effective stress on the rock, where the effective stress is defined as the difference between overburden pressure and pore pressure. Increase in overburden pressure tends to improve the thermal contact between grains and as a result increases thermal conductivity.

Edmondson[16] showed that the thermal conductivities of three sandstones increased by 7·8%, 9·5% and 12·3% per 1000 psi in the 900–3600 psi range. A linear relationship was established between thermal conductivity and pore volume compressibility of the sandstones. Woodside and Messmer,[17] however, obtained a lower increase than Edmondson. Their results were 11·5% per 1000 psi in the range of 0–1000 psi and 2·5% per 1000 psi in the range 2000–4000 psi. We postulate that different cementation and pore geometry characteristics of sandstones may be responsible for the differences.

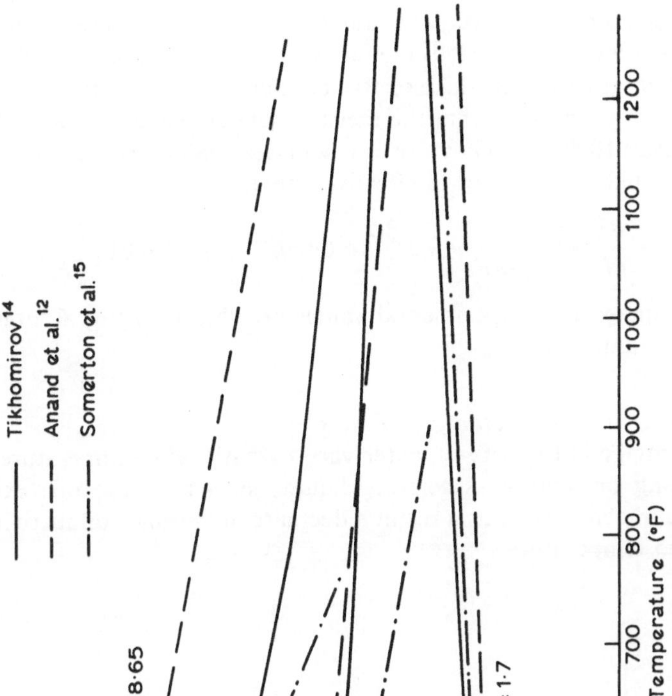

FIG. 5. Comparison of correlations of thermal conductivity with temperature.

If the conductivity of the saturating fluid is increased, the effect of the overburden pressure is decreased and the conductivity of the whole system becomes more uniform. If the effective stress stays constant while pore pressure increases, then the effective conductivity increases.

Anand *et al.*[12] found that the thermal conductivities increased by only 1–2% per 1000 psi. They also calculated values using eqn (2.2.9) differentiated with respect to effective stress:

$$\frac{\partial \lambda}{\partial P} = 0{\cdot}34\frac{\partial \rho}{\partial P} - 3{\cdot}2\frac{\partial \phi}{\partial P} + 0{\cdot}05k^{-0{\cdot}9}\frac{\partial k}{\partial P} + 0{\cdot}013\frac{\partial F}{\partial P} \qquad (2.2.14)$$

The calculated and experimental values are shown in Fig. 6 along with those of Edmondson.[16]

2.2.3. *Thermal Conductivity of Fluids*

The thermal conductivity of water varies greatly with temperature. Non-associating or weakly associating liquids such as crude oils exhibit a decrease in density and as a result a decrease in thermal conductivity with increased temperature.

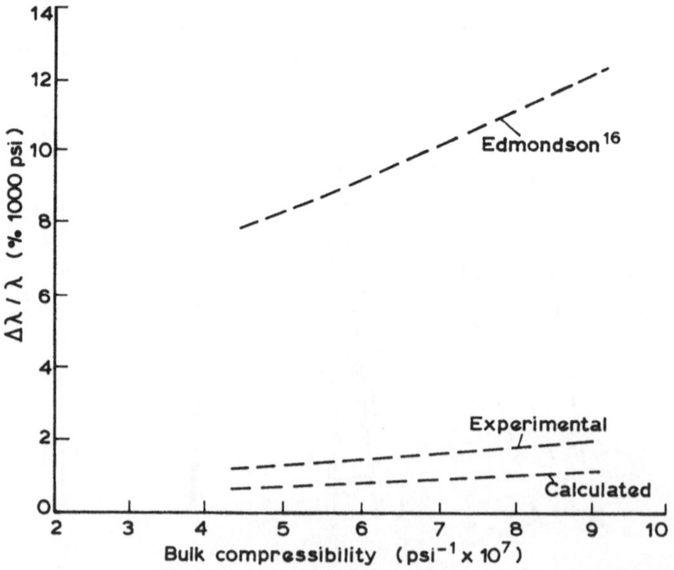

FIG. 6. Effect of pressure on thermal conductivity versus rock compressibility.[12]

Cragoe[18] proposed the following equation for the thermal conductivity of oil fractions and mixtures:

$$\lambda_o = \frac{1{\cdot}62[1 - 0{\cdot}0003(T - 32)]}{\gamma_o} \tag{2.2.15}$$

This equation correlates experimental measurements of oils with an average error of 12% and a maximum error of 39%.

Lenoir[19] defined a thermal conductivity ratio as a function of isothermal expansion factors

$$\frac{\lambda_1}{\lambda_2} = \left(\frac{\omega_1}{\omega_2}\right)^m = \frac{\varepsilon_1}{\varepsilon_2} \tag{2.2.16}$$

where

$$m = -294\frac{T}{T_{cr}} + 3{\cdot}77$$

Values of ε are given in Fig. 7.

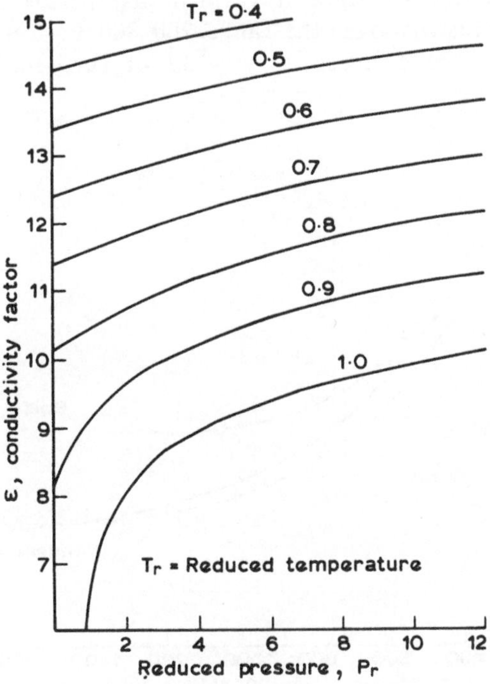

FIG. 7. Generalised chart for liquid thermal conductivities and effect of pressure.[19]

When Lenoir's predicted values are compared with experimental values there is a mean deviation of 1·6%.

Gas thermal conductivity increases with temperature. The kinetic theory of gases combined with an intermolecular potential provides a prediction that thermal conductivity will increase proportionately with the square root of temperature.

2.3. Thermal Diffusivity

Thermal diffusivity is defined as the ratio of thermal conductivity to the product of density and heat capacity:

$$D = \frac{\lambda}{\rho C} \tag{2.3.1}$$

Since increasing temperature results in an increase in heat capacity and a decrease in thermal conductivity, the effect of temperature on diffusivity can be significantly large. A typical variation of thermal conductivity with temperature is shown in Fig. 8. It is evident that there is a marked decrease in thermal diffusivity over the range 200–800 °F. Depending on sand characteristics, the decrease is two-fold or three-fold. Heat transfer

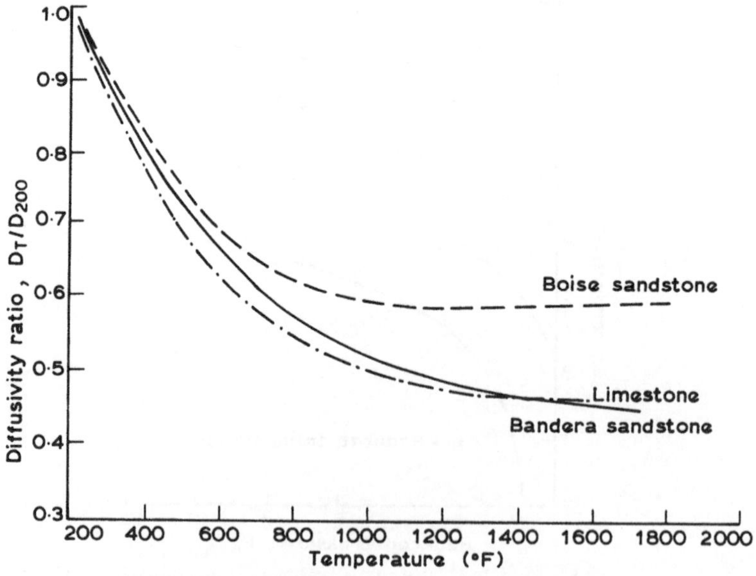

FIG. 8. Thermal diffusivity ratio.[20]

calculations must therefore consider the temperature dependence of diffusivity.[20]

2.4. Pore Volume Compressibility

Laboratory experiments concerning the effect of pressure and temperature on pore volume compressibility have been reported by Von Gonten and Choudhary.[21] Pore volume compressibility may be defined as

$$c_f = -\frac{1}{V_p}\left(\frac{\partial V_p}{\partial P}\right)_T \tag{2.4.1}$$

Figure 9 illustrates the variation of compressibility with effective stress. It is apparent from these results that compressibility decreases as pressure increases to about 6000 psi but beyond that it does not vary much. It is expected that the degree of consolidation and pore cementation will affect the shape of the relationship in natural petroleum reservoir rocks. The effect of temperature on pore volume compressibility for one of the

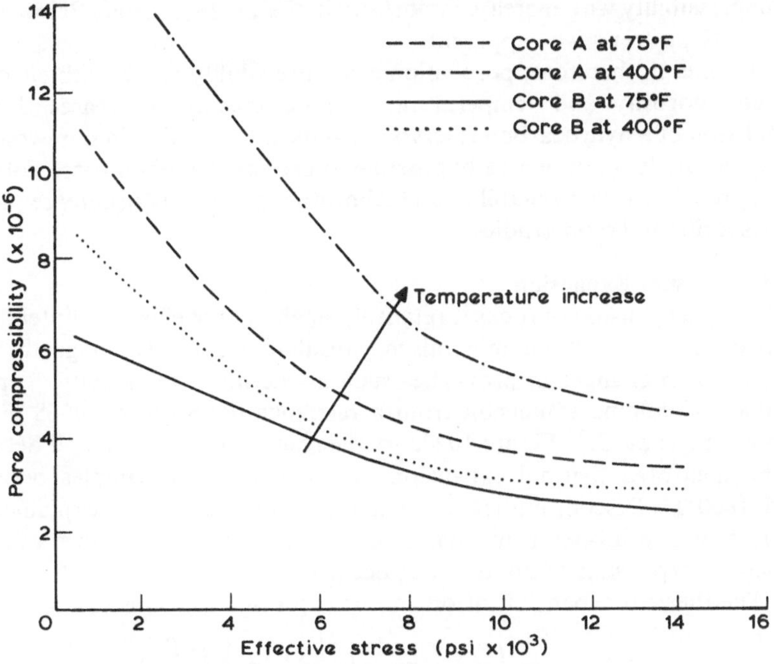

FIG. 9. Pore compressibility (psi^{-1}) versus effective stress.[21]

TABLE 3
VOLUMETRIC THERMAL EXPANSION OF SOLIDS

Material	*Expansion [(%) relative to volume at 68°F] at temperature:*[22]				
	212°F	*392°F*	*752°F*	*1112°F*	*1472°F*
Halite	0·963	2·288	5·256	8·932	
Pyrite	0·219	0·529	1·291		
Aragonite	0·36	1·00	2·48		
Calcite	0·105	0·285	0·765	1·395	
Barite	0·434	1·023	2·381		
Quartz	0·36	0·78	1·87	4·54	4·43
Feldspars	~0·1	~0·3	~0·8	~1·1	~1·5

samples in the data of Ref. 21 suggests that compressibility at 400 °F is about 12 % higher than that at 75 °F. Another sample showed a 55 % variation between 75 and 400 °F. The average of all the samples tested gave a 21 % decrease in this temperature range. The experimental results using sandstone cores at a given effective stress showed an increase in compressibility with increase in porosity in the 19–28 % range. Below 19 % porosity no trend was apparent.

In general, therefore, pore volume compressibility varies with effective stress, porosity and temperature. If pore pressure decreases during depletion of a hydrocarbon reservoir, significant reduction in pore volume may occur. It is not generally possible to predict changes in pore volume compressibility with reliability and laboratory studies are usually required for specific reservoir studies.

2.5. Thermal Expansion

Thermal expansion of rocks is relatively small but the effects of differential expansion of constituent minerals may result in structural damage, thereby resulting in changes in properties such as thermal conductivity. Typical values of volume expansion from a reference temperature of 68 °F are shown in Table 3.[22] Figure 10 shows the results of Somerton and Selim[23] who measured thermal expansion of sandstone core samples between 75–1800 °F. Discontinuities occurring around 1100 °F are explained in terms of α–β inversion in quartz. Over the whole temperature range a thermal expansion of about 1·6 % occurred.

The thermal expansion of oil can be expressed as follows:

$$\beta_o = -\frac{1}{\rho_o}\left(\frac{\partial \rho_o}{\partial T}\right)_P = \frac{1}{V_o}\left(\frac{\partial V_o}{\partial T}\right)_P = \frac{1}{B_o}\left(\frac{\partial B_o}{\partial T}\right)_P \tag{2.5.1}$$

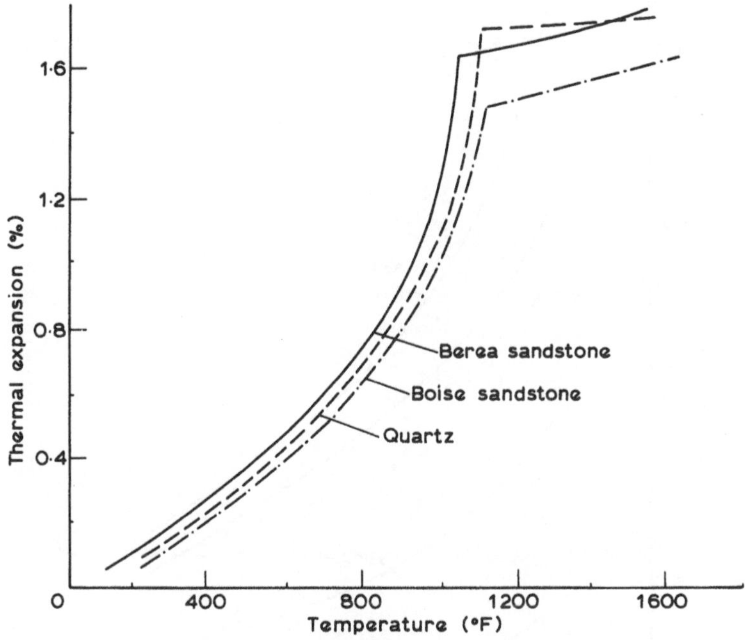

FIG. 10. Thermal expansion versus temperature.[23]

The correlation given by Standing[3] is

$$\rho_{o(ref)} - \rho_o = (0{\cdot}0133 + 152{\cdot}4\rho_{o(ref)}{}^{-2{\cdot}45})(T-60) - [8{\cdot}1(10^{-6}) - 0{\cdot}0622(10^{-0{\cdot}0764}\rho_{o(ref)})](T-60)^2 \qquad (2.5.2)$$

and has a standard deviation of less than 3·4 % for 50 samples. The thermal expansion of water may be evaluated using steam tables.[1,2]

2.6. Viscosity

The effect of viscosity reduction is a key factor in displacement mechanisms in thermal recovery processes. The relative mobilities of fluids in a reservoir vary with temperature and, depending on the various viscosity–temperature relationships, may result in increased oil recovery. Changes in the respective relative permeabilities also influence mobilities but these are also dependent on wetting-phase saturation.

The preferred equation for predicting viscosities at higher temperatures has been found to be that of Andrade; as given by Farouq Ali[24] it is:

$$\mu = A\exp(B/T) \qquad (2.6.1)$$

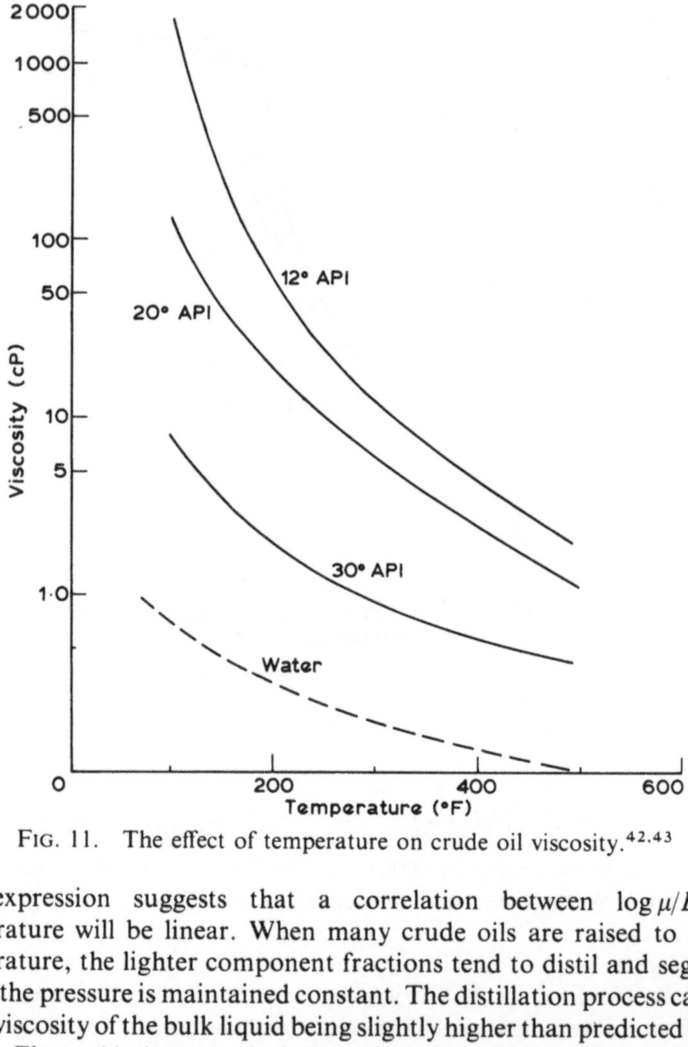

FIG. 11. The effect of temperature on crude oil viscosity.[42,43]

The expression suggests that a correlation between $\log \mu/B$ and temperature will be linear. When many crude oils are raised to a high temperature, the lighter component fractions tend to distil and segregate unless the pressure is maintained constant. The distillation process can lead to the viscosity of the bulk liquid being slightly higher than predicted by eqn (2.6.1). Figure 11 shows typical variations of viscosity with temperature.

3. EFFECT OF TEMPERATURE ON PETROPHYSICAL PROPERTIES

3.1. Absolute Permeability

Several research workers have reported reductions in absolute permeability as temperature is increased. Afinogenov[25] found that absolute permeability decreased sharply, by as much as 88%, with increase of temperature. However he suggested that this decrease could be accounted for by fluid property changes. By definition, absolute permeability is independent of the liquid-phase fluid used to establish the property. With gases, the Klinkenberg correction essentially converts gas flow to equivalent liquid flow. In practice some mineral constituents of the pore space may react to the type of liquid flowing. This is particularly true for clay minerals sensitive to water. The occurrence and effect of illite in pores of North Sea reservoir rocks is particularly sensitive to fluid composition.[26] The results of Weinbrandt *et al.*[27] showed that, for Boise sandstone cores, the absolute permeability fell from 2050 mD at room temperature to 884 mD at 175 °F. This was accounted for by an expansion of the matrix and as a consequence a reduction in porosity as temperature increased.

Other studies[28] have found an increase in permeability with temperature increase. Gobran *et al.*[29] found that absolute permeability to distilled water was independent of temperature for unconsolidated sand and Berea sandstone.

Doubt must be expressed about the general validity of results unless the clay chemistry effects are also considered and the mineralogical composition of the sandstone tested is stated.

3.2. Relative Permeability

As with absolute permeability it is expected that reservoir rock mineralogy is a factor in relative permeability characterisation. It is further expected in multiphase flow that pore geometry is of increased significance. Several studies of the variation in relative permeability with temperature have been made and although the results are not in particular agreement, most report a significant increase in relative permeability to oil as temperature is increased.

Using sandpacks and 80, 99, and 600 cP oil, Poston *et al.*[30] found that both oil and water relative permeabilities increased in a temperature range of 70–300 °F. At a given saturation the relative permeability ratio k_{rw}/k_{ro} increased above that at room temperature but there was no specific trend.

With increasing temperature Weinbrandt and Ramey[27] found that relative permeabilities to water were inconsistent. Increase in water permeability did not occur with increase in temperature for water saturations less than those corresponding to the residual oil saturation.

Edmondson[16] conducted water flooding experiments in the range 75–500 °F and found a variation in relative permeability ratio as temperature increased. Davidson[31] also found the relative permeability ratio to be temperature dependent from 75–540 °F.

Sufi *et al.*[32] have recently studied the effects of temperature on relative permeability up to 122 °F and reported, controversially, that allowing for experimental and computational error, there was no temperature effect. Nakornthap and Evans[33] have derived analytical expressions for relative permeability in terms of water saturation and showed that relative permeability to oil increases as temperature increases. This is shown in Fig. 12. They compared their results with a field example and found that the use of room temperature relative permeabilities in calculations would lead to pessimistic results. This result is well known in reservoir engineering.[34]

In their tests, Miller and Ramey[35] found that there were no changes in relative permeability as temperature was varied. They suggested that previous results were affected by viscous instabilities, end-effects and material balance difficulties.

Maini and Batycky[36] studied the effect of temperature on relative permeabilities in horizontally and vertically drilled core plugs. They found that relative permeability to water increased with temperature. They also found that permeability damage was caused by water-sensitive clays within the cores. Some restoration of high permeability was achieved by injection of a 1 % sodium chloride brine.

In an investigation of the effect of temperature on high and low

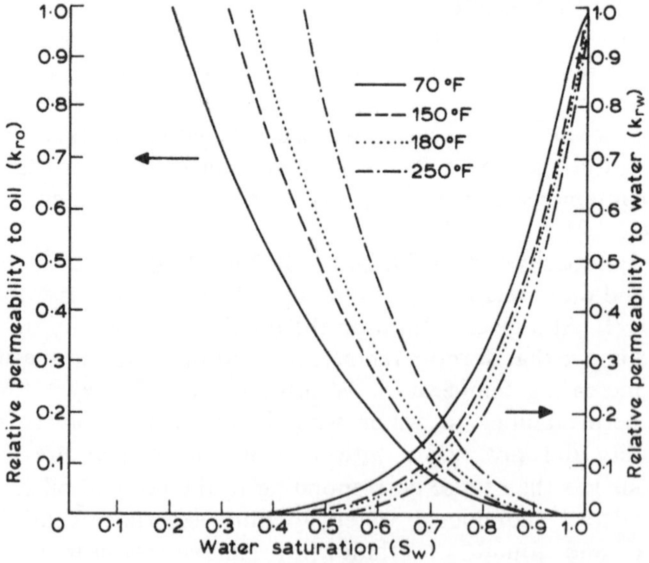

FIG. 12. Temperature dependent relative permeabilities for a field sample.[33]

interfacial tension systems, Torabzadeh and Handy[37] found that in both systems relative permeability to oil increased with temperature. They concluded that the change in wettability of the rock and reduction in interfacial tension were factors involved in the variation of the relative permeability curves.

3.3. Residual Oil Saturation and Irreducible Water Saturation

It is generally agreed that the residual oil saturation, S_{or}, decreases and the irreducible water saturation S_{wi} increases with increasing temperature. The decrease in residual oil saturation is more significant for heavier oils. Poston *et al.*[30] found a decrease from 43 % to 12 % for a 600 cP oil when the temperature was increased from room temperature to 270 °F. Sufi *et al.*[32] found a change in 'practical' residual oil saturation with temperature; as the temperature is increased the viscosity ratio μ_o/μ_w decreases. This results in a change of shape of the fractional flow curves as shown in Fig. 13. It should

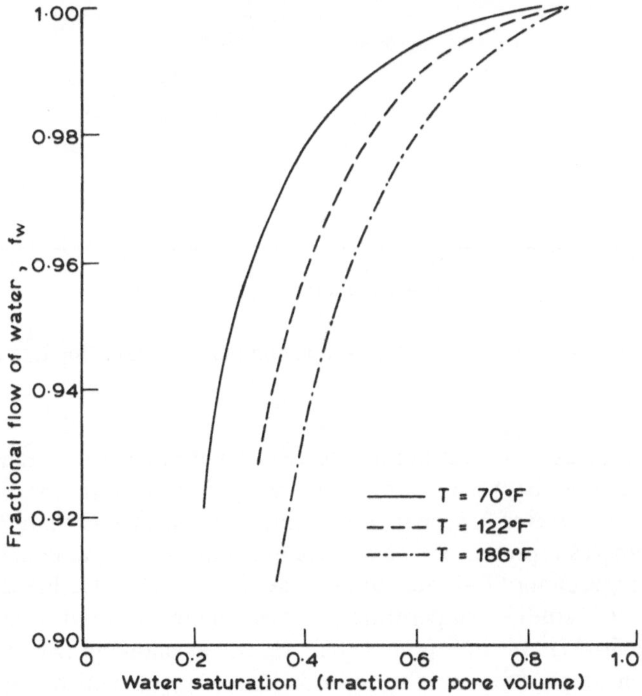

FIG. 13. Water fractional flow versus water saturation for waterfloods at several temperatures.[32]

be noted however that all their curves reach the same end-point, so ultimate residual oil saturation at 100 % fractional flow of water is constant.

Various postulates have been proposed to explain the increase in irreducible saturation as temperature increases. Poston *et al.*[30] and Lo and Mungan[38] found that the oil–water–rock contact angle, θ_c, which might be

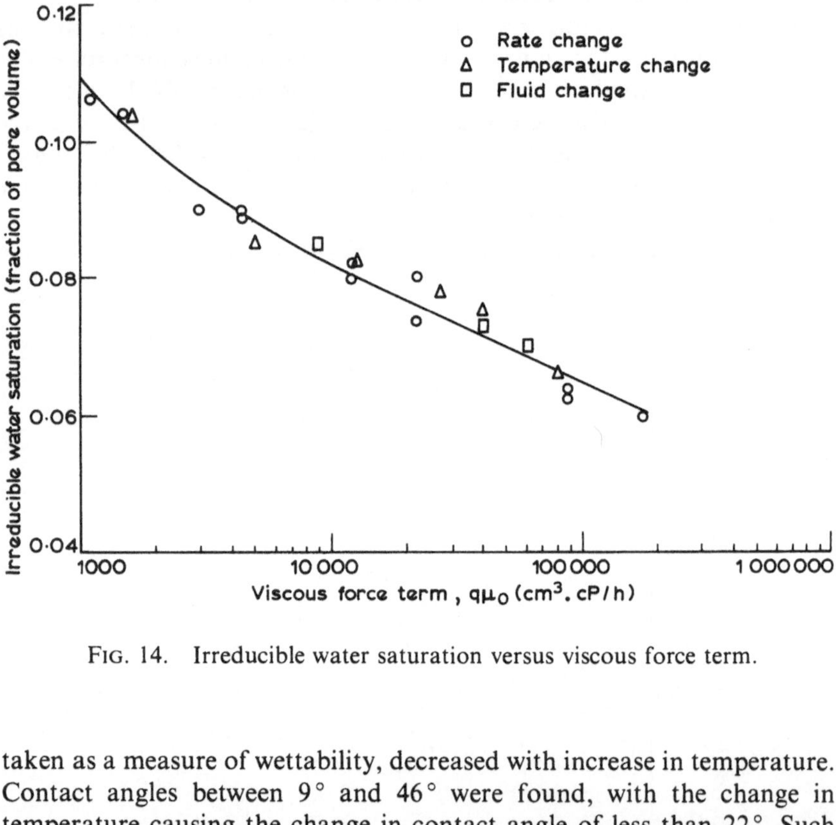

FIG. 14. Irreducible water saturation versus viscous force term.

taken as a measure of wettability, decreased with increase in temperature. Contact angles between 9° and 46° were found, with the change in temperature causing the change in contact angle of less than 22°. Such contact angles represent water wet systems but such small changes in the contact angle cannot fully account for the change in irreducible saturation.

Sufi *et al.*[32] studied temperature effects in oil flow rate and viscosity, and, as shown in Fig. 14, interpreted their results as showing that irreducible saturation depended on the viscous force ($q\mu_o$) which varied with temperature.

3.4. Capillary Pressure

Sanyal *et al.*[39] found that the time needed to reach saturation equilibrium decreased as pressure increased and that the hysteresis between drainage and imbibition curves decreased with increasing temperature. This is shown in Fig. 15. Their reported experimental results were in general

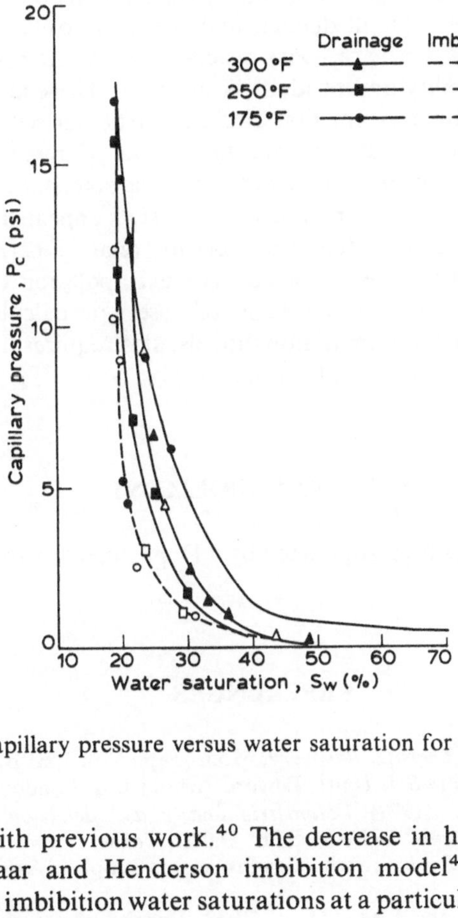

FIG. 15. Capillary pressure versus water saturation for Boise sandstone.[32]

agreement with previous work.[40] The decrease in hysteresis is expected from the Naar and Henderson imbibition model[41] which relates the drainage and imbibition water saturations at a particular capillary pressure to the irreducible water saturation.

$$S_{w,dr} - S_{w,imb} = \frac{(S_{wdr} - S_{wi})^2}{2(1 - S_{wi})} \tag{3.4.1}$$

4. IMPLICATIONS IN RESERVOIR ENGINEERING CALCULATIONS INVOLVING THERMAL PROCESSES

The currently available petroleum literature contains much of the data to define fundamental heat flow characteristics of rock and fluid systems. The accuracy of correlations used to define many of these parameters is, in general, insensitive to results obtained by their use in combined heat and mass balances. The most ill defined and, after viscosity, perhaps most important characteristics in recovery processes, are the temperature effects on relative permeability and residual saturations. These are traditionally empirical and require experimental data for particular reservoir systems. To be relevant they must include the effects of reservoir fluids on constituent minerals in the pore space. The characterisation of reservoir rocks in terms of pore geometry and surface effects appears to be required, in addition to saturation definition, before relative permeability and residual saturations can be predicted. The extrapolation of laboratory-scale displacement data to the scale of reservoir calculations, either analytical or in numerical simulation models, also requires further work to suggest the magnitude of modification.[1,42,43]

ACKNOWLEDGEMENT

Part of this work has been supported by a Department of Energy contract at Imperial College.

REFERENCES

1. Prats, M. (1982) *Thermal Recovery*, Monograph Vol. 7, SPE, Dallas, Texas.
2. *UK Steam Tables in S.I. Units*, Edward Arnold Ltd, London, 1970.
3. Standing, M. B. (1977) *Volumetric and Phase Behaviour of Oil Field Hydrocarbon Systems*, 2nd edn, SPE, Dallas, Texas.
4. *Engineering Data Book*, 9th edn, Gas Processors Suppliers Association, Tulsa, Oklahoma, 1972.
5. Standing, M. B. and Katz, D. L. (1942) Density of natural gases, *Trans. AIME*, **146**, 140–9.
6. Edminster, W. C. (1961) *Applied Hydrocarbon Thermodynamics*, Gulf Publishing Co., Houston, Texas.
7. Somerton, W. H. (1958) Some thermal characteristics of porous rocks. *Trans. AIME*, **213**, 375–8.

8. KESLER, M. G. and LEE, B. I. (1976) Improved prediction of enthalpy of fractions. *Hydrocarbon Processing*, March, 153–8.
9. KUNII, D. and SMITH, J. M. (1961) Thermal conductivities of porous rocks filled with stagnant fluid. *Soc. Pet. Eng. J.*, March, 37–42.
10. ZIERFUSS, H. and VAN DER VLIET, G. (1956) Laboratory measurements of heat conductivity of sedimentary rocks. *Bull. Am. Assoc. Pet. Geologists*, **10**(10), 2475.
11. ASSAD, Y. (1955) A study of the thermal conductivity of fluid bearing porous rocks. PhD Thesis, Univ. of California.
12. ANAND, J., SOMERTON, W. H. and GOMAA, E. (1973) Predicting thermal conductivities of formations from other known properties. *Soc. Pet. Eng. J.*, October, 267–73.
13. EVANS, T. R. (1976) Thermal properties of North Sea rocks. Paper B, *Trans. 4th European Formation Evaluation Symp.*, SPWLA, London.
14. TIKHOMIROV, V. H. (1968) Conductivity of rocks and their relationship with density, saturation and temperature. *Neftianoe Khoziaistro*, **46**(4), 36.
15. SOMERTON, W. H., KEESE, J. A. and CHU, S. L. (1974) Thermal behaviour of unconsolidated oil sands. *Soc. Pet. Eng. J.*, October, 513–21.
16. EDMONDSON, J. A. (1965) Effect of temperature on waterflooding. *J. Can. Pet. Tech.*, **4**, 236–42.
17. WOODSIDE, W. and MESSMER, J. H. (1961) Thermal conductivity of porous media. *J. Appl. Phys.*, **39**(9), 1688.
18. CRAGOE, C. S. (1929) Miscellaneous Publication No. 97. US Bureau of Standards.
19. LENOIR, J. H. (1956) Effect of pressure on thermal conductivity of liquids. *Petroleum Refiner*, **36**(8), 162–4.
20. SOMERTON, W. H. and BOOZER, G. D. (1960) Thermal characteristics of porous rocks at elevated temperature. *Pet. Trans. AIME*, **219**, 418.
21. VON GONTEN, W. D. and CHOUDHARY, B. K. (1969) The effect of pressure and temperature on pore volume compressibility. Paper SPE 2526, *44th Annual Fall Technical Conference*, Denver, Colorado.
22. *Handbook of Physical Constants*, Ed. S. P. Clark Jr (revised edn), Geological Soc. of America, 1966, Memoir 97.
23. SOMERTON, W. H. and SELIM, M. A. (1961) Additional thermal data for porous rocks—thermal expansion and heat of reaction. *Soc. Pet. Eng. J.*, December, 249–53.
24. FAROUQ ALI, S. M. (1967) Estimation, correlation and measurement of oil viscosity at high temperatures. *Producers Monthly*, February, 8.
25. AFINOGENOV, Y. A. (1969) How the liquid permeability of rocks is affected by pressure and temperature. *SNIIGIMS*, No. 6, 34.
26. BROWN, C. E. and LANGLEY, G. O. (1981) The provision of laboratory data for EOR simulation. *1981 European Symposium on Enhanced Oil Recovery*, Elsevier, Amsterdam.
27. WEINBRANDT, R. H. and RAMEY, H. J. (1975) The effect of temperature on relative permeability of consolidated rocks. *Soc. Pet. Eng. J.*, October, 376–84.
28. AKTAN, T. and FAROUQ ALI, S. H. (1975) Effect of cyclic and *in-situ* heating on the absolute permeability, elastic constants, and electrical resistibilities of rocks. Paper SPE 5633, *50th Annual Fall Technical Conference*, Dallas, Texas.

29. Gobran, B. D., Brigham, W. E. and Ramey, H. J. (1981) Absolute permeability as a function of confining pressure, pore pressure and temperature. Paper SPE 10156, *56th Annual Fall Technical Conference*, San Antonio, Texas.
30. Poston, S. N., Ysrael, S., Hassain, A. K. M. S., Montgomery, E. F. and Ramey, H. J. (1970) The effect of temperature on irreducible water saturation and relative permeability of unconsolidated sands. *Soc. Pet. Eng. J.*, **10** (June), 171–80.
31. Davidson, L. B. (1969) The effect of temperature on the permeability ratio of different fluid pairs in two-phase systems. *J. Pet. Tech.*, August, 1037–46.
32. Sufi, A. H., Ramey, H. J. and Brigham, W. E. (1982) Temperature effects on relative permeabilities of oil–water systems. Paper SPE 11071, *57th Annual Fall Technical Conference*, New Orleans, Louisiana.
33. Nakornthap, K. and Evans, R. D. (1982) Temperature dependent relative permeability and its effect upon oil displacement by thermal methods. Paper SPE 1217, *57th Annual Fall Technical Conference*, New Orleans, Louisiana.
34. Colpitts, G. P. and Hunter, D. E. (1964) Laboratory displacement of oil by water under simulated reservoir conditions. *J. Can. Pet. Tech.*, **3**(2), 66–70.
35. Miller, M. A. and Ramey, H. J., Jr. (1983) Effect of temperature on oil/water relative permeabilities of unconsolidated and consolidated sands. Paper SPE 12116, *58th Annual Fall Technical Conference*, San Francisco, California.
36. Maini, B. B. and Batycky, J. P. (1983) Effect of temperature on heavy-oil/water relative permeabilities in horizontally and vertically drilled core plugs. Paper SPE 12115, *58th Annual Fall Technical Conference*, San Francisco, California.
37. Torabzadeh, S. J. and Handy, L. L. (1984) The effect of temperature and interfacial tension on water/oil relative permeabilities of consolidated sands. Paper SPE 12689, *SPE/DoE 4th Symposium on Enhanced Oil Recovery*, Tulsa, Oklahoma.
38. Lo, H. Y. and Mungan, N. (1973) Effect of temperature on water–oil relative permeabilities in oil-wet and water-wet systems. Paper SPE 4505, *48th Annual Fall Technical Conference*, Las Vegas, Nevada.
39. Sanyal, S. K., Ramey, H. J. Jr and Marsden, S. S. (1973) The effect of temperature on capillary pressure properties of rocks. *Trans. 14th Logging Symposium*, Soc. Prof. Well Log Analysts, Lafayette, Indiana.
40. Sinnokrot, A. A., Ramey, H. J. Jr and Marsden, S. S. (1971) Effect of temperature level upon capillary pressure curves. *Soc. Pet. Eng. J.*, March, 13.
41. Naar, J. and Henderson, J. H. (1962) An imbibition model—its application to flow behaviour and the prediction of oil recovery. *Pet. Trans. AIME*, 13.
42. Gates, C. F. and Ramey, H. J. (1964) Better technology opens way for more thermal projects. *Oil and Gas Journal*, July 13th, 72.
43. Dingley, A. J. (1965) The combustion recovery process—principles and practice. *J. Can. Pet. Tech.*, October–December, 196.

INDEX